Lecture Notes in Computer Science   15199

Founding Editors

Gerhard Goos
Juris Hartmanis

Editorial Board Members

Elisa Bertino, *Purdue University, West Lafayette, IN, USA*
Wen Gao, *Peking University, Beijing, China*
Bernhard Steffen ⓘ, *TU Dortmund University, Dortmund, Germany*
Moti Yung ⓘ, *Columbia University, New York, NY, USA*

The series Lecture Notes in Computer Science (LNCS), including its subseries Lecture Notes in Artificial Intelligence (LNAI) and Lecture Notes in Bioinformatics (LNBI), has established itself as a medium for the publication of new developments in computer science and information technology research, teaching, and education.

LNCS enjoys close cooperation with the computer science R & D community, the series counts many renowned academics among its volume editors and paper authors, and collaborates with prestigious societies. Its mission is to serve this international community by providing an invaluable service, mainly focused on the publication of conference and workshop proceedings and postproceedings. LNCS commenced publication in 1973.

Sharib Ali · Fons van der Sommen ·
Bartłomiej Władysław Papież · Noha Ghatwary ·
Yueming Jin · Iris Kolenbrander
Editors

Cancer Prevention, Detection, and Intervention

Third MICCAI Workshop, CaPTion 2024
Held in Conjunction with MICCAI 2024
Marrakesh, Morocco, October 6, 2024
Proceedings

Editors
Sharib Ali
University of Leeds
Leeds, UK

Fons van der Sommen
Eindhoven University of Technology
Eindhoven, The Netherlands

Bartłomiej Władysław Papież
University of Oxford
Oxford, Oxfordshire, UK

Noha Ghatwary
Arab Academy for Science and Technology
Giza, Egypt

Yueming Jin
National University of Singapore
Singapore, Singapore

Iris Kolenbrander
Eindhoven University of Technology
Eindhoven, The Netherlands

ISSN 0302-9743   ISSN 1611-3349 (electronic)
Lecture Notes in Computer Science
ISBN 978-3-031-73375-8   ISBN 978-3-031-73376-5 (eBook)
https://doi.org/10.1007/978-3-031-73376-5

© The Editor(s) (if applicable) and The Author(s), under exclusive license
to Springer Nature Switzerland AG 2025

This work is subject to copyright. All rights are solely and exclusively licensed by the Publisher, whether the whole or part of the material is concerned, specifically the rights of translation, reprinting, reuse of illustrations, recitation, broadcasting, reproduction on microfilms or in any other physical way, and transmission or information storage and retrieval, electronic adaptation, computer software, or by similar or dissimilar methodology now known or hereafter developed.
The use of general descriptive names, registered names, trademarks, service marks, etc. in this publication does not imply, even in the absence of a specific statement, that such names are exempt from the relevant protective laws and regulations and therefore free for general use.
The publisher, the authors and the editors are safe to assume that the advice and information in this book are believed to be true and accurate at the date of publication. Neither the publisher nor the authors or the editors give a warranty, expressed or implied, with respect to the material contained herein or for any errors or omissions that may have been made. The publisher remains neutral with regard to jurisdictional claims in published maps and institutional affiliations.

This Springer imprint is published by the registered company Springer Nature Switzerland AG
The registered company address is: Gewerbestrasse 11, 6330 Cham, Switzerland

If disposing of this product, please recycle the paper.

Preface

CaPTion 2024 was the "3rd International Workshop on Cancer Prevention, detection and intervenTion", organised as a satellite event of the 27th International Conference on Medical Image Computing and Computer Assisted Intervention (MICCAI 2024) in Marrakesh, Morocco. The main idea of founding CaPTion was to create a new research interface where researchers in medical image analysis, machine learning, and clinical research could interact and address cancer-related challenges and early cancer detection using computational methods. In this edition, we also added a new theme on interventional oncology.

Cancer diagnosis and its treatment for the long-term survival of cancer patients have been a battle for decades. 19.3 million cancer cases and almost 10 million deaths were reported in 2020, with lung (18%), colorectal (9.4%), liver (8.3%), stomach (7.7%), and female breast cancer (6.9%) being the leading causes of mortality. While computational methods in medical imaging have enabled us to detect and assess cancerous tumours and assist in their treatment, progress in methodology development with machine learning focussed on cancer understanding can enable timely and optimal diagnosis and interventions critical for cancer prevention.

The survival rate of cancer is still low in the majority of countries and largely depends on the affected organ and how early it is diagnosed. This is critically more important in Africa and Asia, where access to digital technologies for early cancer diagnosis and intervention planning is limited. It was reported that Egypt, Nigeria, South Africa, Ethiopia, and Morocco had the highest incidence and mortality in 2020, accounting for 45% of cases and 44% of cancer deaths. The heterogeneous nature of the cancer disease in different patients and the diverse imaging acquisition types involved in quantifying disease and treatment demand robust method designs. It is, therefore, critical to develop generalisable methods as part of a holistic early cancer detection ecosystem - including different data analysis methods for various modalities and optimising cross-modality fusion for improved detection and prognosis.

The workshop provided an opportunity to present research work in medical imaging around the central theme of cancer detection, prevention, and intervention. It strove to address the challenges that must be overcome to translate computational methods into clinical practice through well-designed, generalisable, interpretable, and clinically transferable methods. Through this workshop, we aimed to identify a new ecosystem that would enable comprehensive method validation and reliability, setting up a new gold standard for sample size and elaborating evaluation strategies to identify failure modes of methods when applied to real-world clinical environments.

The CaPTion 2024 proceedings contain 22 high-quality papers of between 8 and 10 pages pre-selected through a rigorous peer review process from 25 submissions (with an average of three reviews per paper). All submissions were peer reviewed through a double-blind process by at least three members of the scientific review committee, comprising 21 experts (including chairs) in the field of medical imaging.

The accepted manuscripts cover various medical image analysis methods primarily focused on cancer and early cancer detection, progression, inflammation understanding, multimodality data, computer-aided navigation, and computational support for intervention. In addition to the papers presented in this LNCS volume, the workshop included three keynote presentations from world-renowned experts: Kristy K. Brock (University of Texas MD Anderson Cancer Center, USA), Ulaş Bağci (Northwestern University, USA), and Peter Moutney (CEO and founder of Odin Vision, UK).

We wish to thank all the CaPTion 2024 authors for their participation and the members of the scientific review committee for their feedback and commitment to the workshop.

October 2024

Sharib Ali
Fons van der Sommen
Noha Ghatwary
Bartłomiej Władysław Papież
Yueming Jin
Iris Kolenbrander
Raneem Toman
Pedro Chavarrias

Organization

Program Committee Chairs

Sharib Ali — University of Leeds, UK
Fons van der Sommen — Eindhoven University of Technology, Netherlands
Noha Ghatwary — Arab Academy for Science and Technology, Egypt
Bartłomiej W. Papież — University of Oxford, UK
Yueming Jin — University College London, UK
Iris Kolenbrander — Eindhoven University of Technology, Netherlands

Student Representative Board Members

Raneem Toman — University of Leeds, UK
Pedro Chavarrias — University of Leeds, UK

Keynote Speakers

Kristy K. Brock — University of Texas MD Anderson Cancer Center, USA
Ulaş Bağci — Northwestern University, USA
Peter Moutney — Odin Vision, UK

Scientific Review Committee

Pedro Chavarrias — University of Leeds, UK
Cris Claessens — Eindhoven University of Technology, Netherlands
Christian Daul — University of Lorraine, France
Mariia Dmitrieva — Seechange, UK
Ricardo Espinosa — Universidad Panamericana, Mexico
Gerardo Emmanuel Loza Galindo — University of Leeds, UK
Iani Gayo — University College London, UK
Noha Ghatwary — Arab Academy for Science and Technology, Egypt

Sayed Mohammad Mostafavi Isfahani	Lunit Inc., Republic of Korea
Mark Janse	UMC Utrecht, Netherland
Adrian Krenzer	University of Würzburg, Germany
Soumya Kundu	King's College London, UK
Carolus Kusters	Eindhoven University of Technology, Netherlands
Daniel Lang	Helmholtz Center Munich, Germany
Van Linh Le	University of Bordeaux, France
Francisco Lopez-Tiro	Tecnológico de Monterrey, Mexico
Christian Mata	Universidad Politécnica de Catalunya, Spain
Patrice Monkam	Tsinghua University, China
Gilberto Ochoa-Ruiz	Tecnológico de Monterrey, Mexico
Bartłomiej Papież	University of Oxford, UK
Karen Sánchez	Universidad Industrial de Santander, Colombia
Georgia Sovatzidi	University of Thessaly, Greece
Mansoor Ali Teevno	Tecnológico de Monterrey, Mexico
Raneem Toman	University of Leeds, UK
Fons van der Sommen	Eindhoven University of Technology, Netherlands
Christiaan Viviers	Eindhoven University of Technology, Netherlands
Ziang Xu	University of Oxford, UK

Contents

Classification and Characterization

Multi-center Ovarian Tumor Classification Using Hierarchical
Transformer-Based Multiple-Instance Learning 3
   *Cris H.B. Claessens, Eloy W.R. Schultz, Anna Koch, Ingrid Nies,
Terese A.E. Hellström, Joost Nederend, Ilse Niers-Stobbe,
Annemarie Bruining, Jurgen M.J. Piek, Peter H.N. De With,
and Fons van der Sommen*

FoTNet Enables Preoperative Differentiation of Malignant Brain Tumors
with Deep Learning 14
   Chenyi Hong, Hualiang Wang, Zhuoxuan Wu, Zuozhu Liu, and Junhui Lv

Classification of Endoscopy and Video Capsule Images Using
CNN-Transformer Model 26
   *Aliza Subedi, Smriti Regmi, Nisha Regmi, Bhumi Bhusal, Ulas Bagci,
and Debesh Jha*

Multimodal Deep Learning-Based Prediction of Immune Checkpoint
Inhibitor Efficacy in Brain Metastases 37
   *Tobias R. Bodenmann, Nelson Gil, Felix J. Dorfner,
Mason C. Cleveland, Jay B. Patel, Shreyas Bhat Brahmavar,
Melisa S. Guelen, Dagoberto Pulido-Arias,
Jayashree Kalpathy-Cramer, Jean-Philippe Thiran, Bruce R. Rosen,
Elizabeth Gerstner, Albert E. Kim, and Christopher P. Bridge*

Seeing More with Less: Meta-learning and Diffusion Models for Tumor
Characterization in Low-Data Settings 48
   Eva Pachetti and Sara Colantonio

Performance Evaluation of Deep Learning and Transformer Models Using
Multimodal Data for Breast Cancer Classification 59
   *Sadam Hussain, Mansoor Ali, Usman Naseem,
Beatriz Alejandra Bosques Palomo, Mario Alexis Monsivais Molina,
Jorge Alberto Garza Abdala, Daly Betzabeth Avendano Avalos,
Servando Cardona-Huerta, T. Aaron Gulliver,
and Jose Gerardo Tamez Pena*

Detection and Segmentation

On Undesired Emergent Behaviors in Compound Prostate Cancer
Detection Systems 73
   Erlend Sortland Rolfsnes, Philip Thangngat, Trygve Eftestøl,
   Tobias Nordström, Fredrik Jäderling, Martin Eklund,
   and Alvaro Fernandez-Quilez

Optimizing Multi-expert Consensus for Classification and Precise
Localization of Barrett's Neoplasia 83
   Carolus H. J. Kusters, Tim G. W. Boers, Tim J. M. Jaspers,
   Martijn R. Jong, Rixta A. H. van Eijck van Heslinga,
   Albert J. de Groof, Jacques J. Bergman, Fons van der Sommen,
   and Peter H. N. De With

Automated Hepatocellular Carcinoma Analysis in Multi-phase CT
with Deep Learning 93
   Krzysztof Kotowski, Bartosz Machura, Damian Kucharski,
   Benjamín Gutiérrez-Becker, Agata Krason, Jean Tessier,
   and Jakub Nalepa

Refining Deep Learning Segmentation Maps with a Local Thresholding
Approach: Application to Liver Surface Nodularity Quantification in CT 104
   Sisi Yang, Alexandre Bône, Thomas Decaens, and Joan Alexis Glaunes

Uncertainty-Aware Deep Learning Classification for MRI-Based Prostate
Cancer Detection 114
   Kamilia Taguelmimt, Hong-Phuong Dang, Gustavo Andrade Miranda,
   Dimitris Visvikis, Bernard Malavaud, and Julien Bert

Generalized Polyp Detection from Colonoscopy Frames Using Proposed
EDF-YOLO8 Network 124
   Alyaa Amer, Alaa Hussein, Noushin Ahmadvand, Sahar Magdy,
   Abas Abdi, Nasim Dadashi Serej, Noha Ghatwary, and Neda Azarmehr

AI-Assisted Laryngeal Examination System 133
   Chiara Baldini, Muhammad Adeel Azam, Madelaine Thorniley,
   Claudio Sampieri, Alessandro Ioppi, Giorgio Peretti,
   and Leonardo S. Mattos

UltraWeak: Enhancing Breast Ultrasound Cancer Detection
with Deformable DETR and Weak Supervision 144
   Ufaq Khan, Umair Nawaz, and Abdulmotaleb E. Saddik

SelectiveKD: A Semi-supervised Framework for Cancer Detection in DBT
Through Knowledge Distillation and Pseudo-labeling 154
   *Laurent Dillard, Hyeonsoo Lee, Weonsuk Lee, Tae Soo Kim, Ali Diba,
and Thijs Kooi*

Cancer/Early Cancer Detection, Treatment, and Survival Prognosis

AI Age Discrepancy: A Novel Parameter for Frailty Assessment in Kidney
Tumor Patients 167
   *Rikhil Seshadri, Jayant Siva, Angelica Bartholomew, Clara Goebel,
Gabriel Wallerstein-King, Beatriz López Morato, Nicholas Heller,
Jason Scovell, Rebecca Campbell, Andrew Wood, Michal Ozery-Flato,
Vesna Barros, Maria Gabrani, Michal Rosen-Zvi, Resha Tejpaul,
Vidhyalakshmi Ramesh, Nikolaos Papanikolopoulos, Subodh Regmi,
Ryan Ward, Robert Abouassaly, Steven C. Campbell, Erick Remer,
and Christopher Weight*

Deep Neural Networks for Predicting Recurrence and Survival in Patients
with Esophageal Cancer After Surgery 176
   *Yuhan Zheng, Jessie A. Elliott, John V. Reynolds, Sheraz R. Markar,
Bartłomiej W. Papież, and ENSURE study group*

Treatment Efficacy Prediction of Focused Ultrasound Therapies Using
Multi-parametric Magnetic Resonance Imaging 190
   *Amanpreet Singh, Samuel Adams-Tew, Sara Johnson, Henrik Odeen,
Jill Shea, Audrey Johnson, Lorena Day, Alissa Pessin, Allison Payne,
and Sarang Joshi*

SurRecNet: A Multi-task Model with Integrating MRI and Diagnostic
Descriptions for Rectal Cancer Survival Analysis 200
   *Runqi Meng, Zonglin Liu, Yiqun Sun, Dengqiang Jia, Lin Teng,
Qiong Ma, Tong Tong, Kaicong Sun, and Dinggang Shen*

Improved Prediction of Recurrence After Prostate Cancer Radiotherapy
Using Multimodal Data and *in Silico* simulations 211
   *Valentin Septiers, Carlos Sosa-Marrero, Renaud de Crevoisier,
Aurélien Briens, Hilda Chourak, Maria A. Zuluaga, and Oscar Acosta*

AutoDoseRank: Automated Dosimetry-Informed Segmentation Ranking
for Radiotherapy 221
   *Zahira Mercado, Amith Kamath, Robert Poel, Jonas Willmann,
Ekin Ermis, Elena Riggenbach, Lucas Mose, Nicolaus Andratschke,
and Mauricio Reyes*

SurvCORN: Survival Analysis with Conditional Ordinal Ranking Neural
Network 231
  *Muhammad Ridzuan, Numan Saeed, Fadillah Adamsyah Maani,
  Karthik Nandakumar, and Mohammad Yaqub*

Author Index 241

Classification and Characterization

Multi-center Ovarian Tumor Classification Using Hierarchical Transformer-Based Multiple-Instance Learning

Cris H.B. Claessens[1(✉)], Eloy W.R. Schultz[1], Anna Koch[2], Ingrid Nies[2], Terese A.E. Hellström[1], Joost Nederend[3], Ilse Niers-Stobbe[4], Annemarie Bruining[5], Jurgen M.J. Piek[2], Peter H.N. De With[1], and Fons van der Sommen[1]

[1] Department of Electrical Engineering, Group of Video Coding and Architectures, Eindhoven University of Technology, Eindhoven, The Netherlands
{c.h.b.claessens,p.h.n.de.with,fvdsommen}@tue.nl
[2] Department of Obstetrics and Gynaecology and Catharina Cancer Institute, Catharina Hospital, Eindhoven, The Netherlands
[3] Department of Radiology and Catharina Cancer Institute, Catharina Hospital, Eindhoven, The Netherlands
[4] Department of Radiology, Amphia Hospital, Breda, The Netherlands
[5] Department of Radiology, Dutch Cancer Institute - Antoni van Leeuwenhoek Hospital, Amsterdam, The Netherlands

Abstract. Malignant ovarian tumors (OTs) are a leading cause of gynecological cancer deaths, and often remain asymptomatic until advanced stages, making early and accurate diagnosis crucial for effective treatment and good patient outcome. Current diagnostic methods often fall short due to the heterogeneous nature of OTs and the complexities in distinguishing benign from malignant forms. To overcome these limitations, this study proposes a novel framework leveraging transformer-based multiple-instance learning (MIL) and hierarchical self-supervised pretraining. To validate the model, a comprehensive multi-center dataset has been compiled, encompassing diverse patient demographics and imaging protocols. Benchmarking against conventional radiomics methods and other deep learning approaches, the hierarchical MIL model demonstrates superior performance with a median AUROC of 0.84 and high recall of 0.91. These results highlight significant improvements in sensitivity, essential for minimizing false negatives in clinical settings. The performed study emphasizes the importance of multi-center validation and external dataset testing to ensure generalization of the proposed model and obtain a higher robustness. The encountered complexity of multi-center data is found significant, since various clinical factors play an influential role. This makes baseline comparisons virtually impossible and the need for more multi-center research increasingly compelling and encouraging.

Keywords: ovarian cancer · computer-aided diagnosis · self-supervised learning · multiple-instance learning · vision transformer · external validation

1 Introduction

Ovarian cancer is a leading cause of gynecological cancer deaths, and often remains asymptomatic until advanced stages, making early and accurate diagnosis crucial for effective treatment and good patient outcome. However, ovarian tumors (OTs) present a significant clinical challenge due to their heterogeneous nature and the difficulty in distinguishing benign from malignant forms. Conventional diagnostic methods, such as ultrasound and serum biomarkers, frequently lack sufficient sensitivity and specificity and are heavily reliant on the operator's expertise [15,16]. For OTs, achieving high sensitivity is particularly crucial as false negatives can have severe consequences for the patient, such as additional surgery and significant physiological burden [6,13,21]. Although computed tomography (CT) imaging provides detailed insights into the anatomical and morphological characteristics of ovarian masses, its accurate interpretation demands significant expertise and can be subjective, leading to variation in diagnosis [19].

Ultrasound is typically the primary imaging modality for diagnosing ovarian tumors, as it is non-invasive and widely available. However, patients who undergo CT are often pre-selected based on suspicious findings from initial ultrasound examinations, which can introduce selection bias. This is because patients with benign tumors often do not receive CT scans, while those with findings suggestive of malignancy or complications are more likely to undergo further evaluation with CT to assess disease extent rather than to make a primary diagnosis [20].

In recent years, computer-aided diagnosis (CADx) has emerged as a promising approach to enhance the accuracy and consistency of OT classification based on CT imagery. Significant advances in machine learning, particularly radiomics and deep learning, have enabled these techniques to automatically extract and analyze features from imaging data, aiming to differentiate benign from malignant tumors [10–12,18]. Despite their potential, many existing studies have considerable limitations, such as reliance on validation cohorts without appropriate test sets, leading to over-fitting and overly optimistic performance estimates [9]. This test-set issue is particularly relevant for OTs due to their high heterogeneity [1,4]. Furthermore, the lack of validation using data from different medical centers raises concerns about the generalization and robustness of these models. A model that performs well on internal training and tuning data has difficulties with generalization to new data without external validation of data from other institutions, due to variations in (a) imaging protocols, (b) patient demographics, and (c) disease presentation. For achieving high performance and robustness with CADx systems, it is required to address the inherent heterogeneity of OT classification and the quality variation due to multi-center data usage. These aforementioned aspects can severely limit the clinical application and reliability of CADx systems.

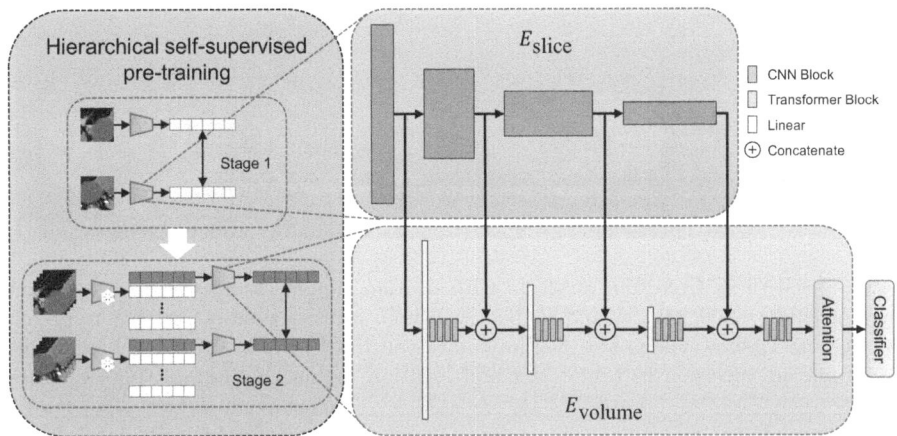

Fig. 1. Overview of the proposed framework. Left: hierarchical self-supervised pre-training divided in two stages. Right: magnified views of the two encoding models within the two stages. Encoder E_{slice} is designed to extract meaningful features from axial data slices of tumors. Encoder E_{volume} integrates these features into a comprehensive volumetric data representation of the entire tumor using a multiple-instance learning approach.

To address these challenges, we have compiled a comprehensive multi-center dataset, incorporating data from several institutions to ensure diverse and representative training and validation cohorts. This dataset includes a wide range of OT cases, capturing variations in imaging hardware and protocols, patient demographics, and disease presentations, which is crucial for developing robust and generalizable models. Borderline OTs, which exhibit features between benign and malignant, were specifically left out of consideration within this work to avoid ambiguity.

Using this dataset, we benchmark several baseline models, including conventional radiomics and various deep learning architectures, to establish a performance baseline. These models are evaluated using nested cross-validation on the multi-center dataset to rigorously assess their performance and mitigate overfitting. Additionally, we evaluate the models against an external dataset from another medical center to further validate their generalization and robustness to show the potential risks of these models when applied to real-world data.

Building on these benchmarks, we propose a novel model based on transformer-based multiple-instance learning (MIL) and a hierarchical self-supervised pre-training scheme. The proposed model leverages the power of transformers to handle the complex and heterogeneous nature of CT imaging data. Additionally, we introduce a hierarchical pre-training schedule, designed to progressively train the model on increasingly complex tasks, which enhances its ability to generalize across different multi-center datasets. The code and pre-trained model weights are publicly available at www.github.com/cclaess/ovacadx.

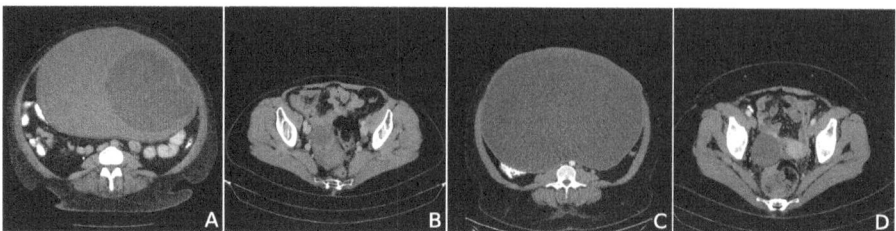

Fig. 2. Axial slices from four subjects with ovarian tumors, windowed to a common range of [-100, +300] HU. A and B display benign ovarian tumors outlined in green, while C and D show malignant ovarian tumors outlined in red. The images highlight the significant variation in tumor size and location within the dataset. (Color figure online)

2 Method

The proposed framework consists of two main encoding components for modeling: a slice encoder (E_{slice}) and a volumetric encoder (E_{volume}). Encoder E_{slice} processes individual CT slices into high-dimensional embeddings that capture rich, hierarchical features. The volume encoder E_{volume}, utilizes a vision transformer architecture and aggregates these slice embeddings to produce a single prediction per volume, following a multiple-instance learning paradigm. To enhance model performance and robustness, we employ a hierarchical self-supervised pre-training scheme. This strategy allows the encoders to learn meaningful representations from unlabeled data, effectively capturing both slice-level features (Stage 1) and volume-level features (Stage 2). The following sections detail the transformer-based multiple-instance learning approach and the hierarchical self-supervised pre-training methodology.

2.1 Transformer-Based Multiple-instance Learning

OTs often grow to considerable size prior to being discovered [1]. Directly extracting features from the obtained CT volumes can pose significant computational and memory challenges. Figure 2 visualizes tumor instances showing the huge spread in size within OTs. This huge spread is considered by clinicians as follows. First, in clinical practice, expert radiologists routinely examine CT scan slices individually during the diagnostic process. Second, the diagnosis typically relies on one or only a few slices adequately representing the entire tumor. As a consequence for our modeling, the large variation leads to a classification problem, which is redefined into a slice-based MIL problem. MIL is suited for this problem because it can learn the individual slices first as instances of one tumor, which are then in later stages combined in one OT model.

Within the MIL framework, the CT volume is treated as a bag $\underline{\underline{B}}$ comprising N instances with $N \in \mathbb{N}$. Each slice is encoded using E_{slice}, yielding N embeddings mapped in feature vector $\boldsymbol{v}$. During training, the input to E_{slice} has a dimensionality of $(N_B * N) \times 1 \times H \times W$, where N_B is the batch size, while H and W represent the height and width of each slice, respectively.

These embeddings $\{\boldsymbol{v_1}, \boldsymbol{v_2}, ..., \boldsymbol{v_N}\}$ are aggregated into a representation $\underline{B} = \{\boldsymbol{v_1}, \boldsymbol{v_2}, ..., \boldsymbol{v_N}\}$ of size $N_B \times N \times V$, with V denoting the embedding size. This aggregated representation is then processed through E_{volume}, which transforms it into a token sequence. Each token corresponds to an axial slice of the volume. Following Myronenko et al. [17], the embeddings $\underline{\underline{B}}$ are fused across different scales to integrate various levels of semantic information (note the different concatenation steps within the encoders in Fig. 1).

2.2 Hierarchical Self-supervised Pre-training

During the self-supervised pre-training phase, the objective is to capture relevant features from both CT slices within E_{slice} and bag embeddings within E_{volume}. We draw inspiration from the approach regularly used for large histopathology images, which employs hierarchical image pyramids [3]. Similarly, we formulate our pre-training task by establishing two hierarchical momentum-distillation frameworks, depicted on the left side of Fig. 1, segmented into Stages 1 and 2.

Initially, we set up a student-teacher architecture and apply a non-contrastive loss, as proposed in Caron et al. [2], to extract meaningful representations from CT slices (Figure 1, Stage 1). This step focuses on learning robust slice-level features. Subsequently, we freeze the encoder weights in E_{slice}, and move to the next hierarchical level. Here, we employ a similar parallel framework on the weights of E_{volume} (Figure 1, Stage 2). The non-contrastive loss from Grill et al. [7] is employed to obtain meaningful representations at the volume level, as the loss proposed by Caron et al. is too memory intensive for a 3D objective.

3 Experiments

3.1 Datasets

In our experiments, we have utilized two distinct datasets to facilitate the pre-training and downstream training phases. To the best of our knowledge, there are no public benchmark datasets specifically focusing on ovarian tumors in CT scans. This absence necessitated the collection and use of private datasets for the downstream training and evaluation of the architectures discussed within this study.

For pre-training, we use the AbdomenCT-1K open-source dataset, comprising 1,062 abdominal CT scans encompassing a variety of pathologies and sourced from multiple medical centers [14]. The diversities in patient demographics and pathologies, along with the multi-center origin of the data, ensure a rich and varied dataset, allowing the models to capture a broad spectrum of features relevant to abdominal CT imaging.

For downstream training and analysis, we have collected a multi-center dataset, including pre-operative CT scans from patients diagnosed with one or two benign or malignant OTs. Final histopathology of these tumors served as the gold standard. The data have been gathered from the Catharina Hospital Eindhoven, The Netherlands, and the Amphia Hospital Breda, The Netherlands. This dataset comprises a total of 370 tumors from 323 subjects, with 115 tumors identified as malignant and 255 as benign. Bilateral tumors from the same subject are always kept within the same split during training. Additionally, we have collected an external test set from The Dutch Cancer Institute - Antoni van Leeuwenhoek Hospital Amsterdam, The Netherlands, comprising 9 benign and 21 malignant OTs.

Both datasets are normalized by windowing between Hounsfield Units (HUs) of -100 to +300 and resampling to a unity range. The CT scans in the collected set are cropped around the tumor regions, and the surrounding tissue is set to background using manually obtained segmented instances provided by an expert radiologist. Examples of these segmentation instances are visualized in Fig. 2.

3.2 Experimental Details

Self-supervised Pre-training: During the self-supervised pre-training phase, both E_{slice} and E_{volume} are trained for 100 epochs using a hierarchical pre-training strategy. The training process employs the Adam optimizer with initial learning rates of 5×10^{-4} for E_{slice} and 3×10^{-4} for E_{volume}. A cosine annealing learning-rate scheduler with a linear warm-up over the first 10 epochs is used. The training was performed using $N_B = 32$ for E_{slice} and $N_B = 8$ for E_{volume}, employing distributed training across two GeForce RTX 3090 Ti GPUs (NVIDIA Corp., CA, USA) with a combined memory of 48 GB.

For E_{slice}, global and local views of axial slices are generated using randomly resized cropping with scale factor $s \in [0.5, 1.0]$ for global views and $s \in [0.125, 0.5]$ for local views. For E_{volume}, two views with $s \in [0.125, 1.0]$ are cropped from the original volume.

Spatial augmentations include random rotations α around the z-axis by $\alpha = \pm 30°$ and random flipping around all axes, following earlier work [5]. Additionally, random Gaussian smoothing or sharpening is applied to simulate different reconstruction kernels.

Ovarian Tumor Classification: For the downstream task of OT classification, the entire framework is trained for an additional 100 epochs, which is sufficient for convergence. Given the high variance within OTs, we employ a nested cross-validation approach to accurately reflect the tumor occurrence frequency in the experimental results. The distribution of the occurrences is such that each subset is stratified between benign and malignant.

This approach includes five datafolds over the test set and five inner datafolds for hyperparameter tuning. The nesting means effectively that training is performed in five major cycles over different test datafolds, and that each time in each cycle the testing data is reused five cycles for hyperparameter tuning on the five inner datafolds.

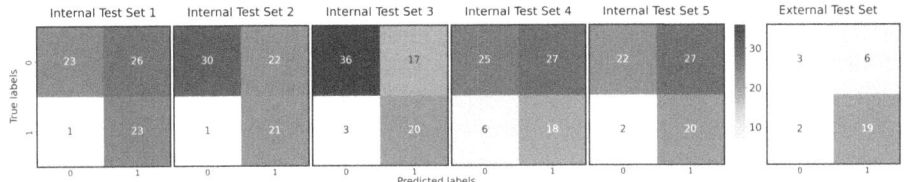

Fig. 3. Confusion matrices of the proposed model for the five test datafolds of the nested cross-validation along with predictions for the external test set.

Table 1. OT classification performance of the evaluated machine-learning and deep learning approaches. Metrics shown with median (min, max) AUROC, precision and recall over the internal test set with nested cross-validation.

Architecture	AUROC ↑	Precision ↑	Recall ↑
Radiomics + LR [10]	0.78 (0.77, 0.82)	0.54 (0.49, 0.57)	0.68 (0.63, 0.83)
Radiomics + RF [10]	0.80 (0.77, 0.84)	**0.62 (0.56, 0.63)**	0.65 (0.54, 0.79)
Radiomics + SVM [10]	0.78 (0.75, 0.82)	0.61 (0.55, 0.67)	0.67 (0.61, 0.79)
Radiomics + MLP [10]	0.79 (0.77, 0.80)	0.58 (0.50, 0.59)	0.71 (0.65, 0.79)
Attention MIL [8]	0.41 (0.36, 0.46)	0.16 (0.11, 0.25)	0.14 (0.08, 0.27)
Transformer MIL [17]	0.68 (0.58, 0.78)	0.43 (0.32, 0.47)	0.68 (0.50, 0.75)
Transformer Pyramid MIL [17]	0.82 (0.72, 0.85)	0.55 (0.44, 0.57)	0.73 (0.67, 0.88)
Proposed model	**0.84 (0.65, 0.89)**	0.47 (0.40, 0.54)	**0.91 (0.75, 0.96)**

During this phase, the weights of E_{slice} are kept constant, while the weights of E_{volume} and the linear classifier are optimized using the Adam optimizer with a base learning rate of 1×10^{-6} and a cosine annealing learning-rate scheduler. The training was conducted using distributed methods on the same hardware as the pre-training, with $N_B = 24$.

We set $N=25$ by *randomly* selecting axial slices covering the tumor. This ensures an even number of instances per sample and prevents excessive memory usage for very large tumors, under the assumption that we are still adhering to the principles of multiple-instance learning. For validation, we *uniformly* sample an equal amount of slices in the same way to obtain a more representative view of the tumor and facilitate fair comparisons between hyperparameter-tuning results. During testing, we evaluate *all* slices covering tumor tissue, making full use of the available data to achieve the most accurate assessment.

Model selection is based on the highest area under the receiver operating characteristic curve (AUROC) on the validation set. For evaluation, precision and recall are also calculated, with the decision threshold optimized to maximize the F_1 score on the validation sets.

The model performance is compared against several benchmarks. We implement the radiomics approach by Li et al. [10] and combine it with various classifiers commonly applied in radiomics, i.e. linear regression (LR), random forest (RF), support vector machine (SVM), and multi-layer perceptron (MLP). Additionally, the proposed model is compared with three MIL frameworks: attention-based MIL [8], transformer-based MIL [17], and pyramid transformer-based MIL [17]. For a fair comparison, we maintain the same training parameters for all benchmarks and use the identical pre-trained weights of E_{slice} for the backbones of these MIL frameworks.

3.3 Results and Discussion

We have compiled a multi-center dataset featuring a diverse set of OTs and compared various radiomics and deep learning models for OT classification. Additionally, we have proposed a novel model based on transformer-based multiple-instance learning and hierarchical self-supervised pre-training. The results of these experiments are summarized in Table 1.

The performance metrics of the radiomics-based approaches highlight the complexity of the encountered data variation in multi-center evaluation, with an AUROC score of 0.79 for the radiomics + MLP approach, compared to an AUROC of 0.91 reported on the validation set in the original paper by Li et al. [10]. This comparison in largely invalid since our datasets contain significant variations in patients, imaging protocols, equipment setting and usage, and expert decisions. Given the involvement of these factors throughout our datasets, the setback of 0.12 in AUROC is fully expected.

Comparison of Radiomics and Deep Learning Approaches: As can be observed in Table 1, radiomics-based models show consistent performance with AUROCs ranging between 0.78–0.80. This consistency suggests robust generalization across different data subsets, as is evidenced by the relatively low spread in the measured performance metrics. Furthermore, radiomics methods achieve a balanced trade-off between sensitivity and specificity, resulting in about the same amount of false positives and false negatives.

In contrast, deep learning approaches exhibit significant variability in performance. The attention-based MIL method completely fails to differentiate between the benign and malignant class, suggesting that the attention mechanism cannot integrate the complex relationships between slices for accurate malignancy prediction. Although the transformer-based MIL approach outperforms the attention-based MIL, it still falls short compared to radiomics in terms of the performance metrics, with a 13–15% decrease in AUROC. Conversely, the transformer pyramid MIL approach exceeds the radiomics models, demonstrating high median AUROC of 0.82 and recall of 0.73. This is probably caused by the increased complexity of the pyramid architecture, allowing the model to capture relationships between slices at different scales. However, it should be noted that the deep learning approaches show a larger variance as compared to radiomics models, thereby suggesting inferior generalization to challenging or less frequent data samples.

Performance of the Proposed Model The proposed model demonstrates superior performance with a median AUROC of 0.84, outperforming all benchmark models. Additionally, we achieve a high recall of 0.91, thereby improving with 28–40% and 25% as compared to radiomics and pyramid transformer-based MIL approaches, respectively. The high recall indicates a low false negative rate, which is crucial for early diagnosis in clinical settings. However, this high recall comes at the cost of a decreased precision, reflecting a higher rate of false positives. This trade-off is crucial in medical diagnosis, where the cost of false negatives (missed malignancies) is much higher than false positives. The confusion matrices of the model (see Fig. 3) show an almost negligible number of false negatives, highlighting its high sensitivity. Nevertheless, even in the best fold, about 33% of negatives are incorrectly predicted as positives, indicating need for improved specificity.

The external test set includes a limited number of benign tumors, limiting reliable qualitative evaluation in terms of AUROC, precision, and recall. Despite this limitation, the confusion matrix at the right side of Fig. 3 shows that our model maintains very high sensitivity even with external data. These results suggest that CADx is a promising tool for OT classification.

It is important to note that the framework used in this study is provided "as is". No experiments were conducted to evaluate the influence of different feature extractors. This decision was driven by the fact that hierarchical pre-training must be repeated for each feature extractor, which is computationally intensive and time-consuming. The primary goal of this paper was to highlight the differences between radiomics and deep learning in the context of multi-center data for ovarian tumor classification and to propose potential solutions like a hierarchical pre-training scheme. Our purpose was not to fully optimize the framework but to provide valuable insights for future research directions.

4 Conclusions

The proposed hierarchical MIL model has shown promise with an AUROC of 0.84 and a high recall of 0.91, significantly outperforming both conventional radiomics and deep learning approaches in classification of OTs and minimizing false negatives. This capability positions the proposed model as an effective tool for early classification of ovarian tumors. Additionally, our study highlights the complexity of multi-center datasets, where multiple influencing factors are variable, and the importance of external validation for developing generalizable models with higher robustness. The introduction of the proposed model marks a significant step forward, but further refinement and validation are required to ensure clinical applicability. Also, we hope that multi-center evaluations become more common practice in research for improved baseline benchmarking.

Acknowledgements. We gratefully acknowledge the Catharina Hospital Eindhoven, the Amphia Hospital Breda, and The Dutch Cancer Institute - Antoni van Leeuwenhoek Hospital Amsterdam for their invaluable data collection support essential to this project.

Disclosure of Interests. The authors have no competing interests to declare that are relevant to the content of this article.

References

1. WHO Ovary Tumor classification. http://www.pathologyoutlines.com/topic/ovarytumorwhoclassif.html
2. Caron, M., Touvron, H., Misra, I., Jégou, H., Mairal, J., Bojanowski, P., Joulin, A.: Emerging Properties in Self-Supervised Vision Transformers (2021). https://doi.org/10.48550/ARXIV.2104.14294, https://arxiv.org/abs/2104.14294, publisher: arXiv Version Number: 2
3. Chen, R.J., et al.: Scaling Vision Transformers to Gigapixel Images via Hierarchical Self-Supervised Learning, pp. 16144–16155 (2022). https://openaccess.thecvf.com/content/CVPR2022/html/Chen_Scaling_Vision_Transformers_to_Gigapixel_Images_via_Hierarchical_Self-Supervised_Learning_CVPR_2022_paper.html
4. Cho, K.R., Shih, I.M.: Ovarian cancer. Ann. Rev. Pathol. **4**, 287–313 (2009). https://doi.org/10.1146/annurev.pathol.4.110807.092246
5. Claessens, C.H.B., et al.: Evaluating task-specific augmentations in self-supervised pre-training for 3D medical image analysis. In: Medical Imaging 2024: Image Processing, vol. 12926, pp. 403–410. SPIE, April 2024. https://doi.org/10.1117/12.3000850. https://www.spiedigitallibrary.org/conference-proceedings-of-spie/12926/129261L/Evaluating-task-specific-augmentations-in-self-supervised-pre-training-for/10.1117/12.3000850.full
6. Geomini, P.M.A.J., Kruitwagen, R.F.P.M., Bremer, G.L., Massuger, L., Mol, B.W.J.: Should we centralise care for the patient suspected of having ovarian malignancy? Gynecol. Oncol. **122**(1), 95–99 (2011). https://doi.org/10.1016/j.ygyno.2011.03.005. https://www.sciencedirect.com/science/article/pii/S0090825811001739
7. Grill, J.B., et al.: Bootstrap your own latent: a new approach to self-supervised Learning (2020). https://doi.org/10.48550/ARXIV.2006.07733. https://arxiv.org/abs/2006.07733, publisher: arXiv Version Number: 3
8. Ilse, M., Tomczak, J.M., Welling, M.: Attention-based Deep Multiple Instance Learning (2018). https://doi.org/10.48550/ARXIV.1802.04712. https://arxiv.org/abs/1802.04712, publisher: [object Object] Version Number: 4
9. Koch, A.H., et al.: Analysis of computer-aided diagnostics in the preoperative diagnosis of ovarian cancer: a systematic review. Insights Imaging **14**(1), 34 (2023). https://doi.org/10.1186/s13244-022-01345-x. https://doi.org/10.1186/s13244-022-01345-x
10. Li, J., Zhang, T., Ma, J., Zhang, N., Zhang, Z., Ye, Z.: Machine-learning-based contrast-enhanced computed tomography radiomic analysis for categorization of ovarian tumors. Front. Oncol. **12**, 934735 (2022). https://doi.org/10.3389/fonc.2022.934735

11. Li, S., et al.: A radiomics approach for automated diagnosis of ovarian neoplasm malignancy in computed tomography. Sci. Rep. **11**, 8730 (2021). https://doi.org/10.1038/s41598-021-87775-x
12. Liu, P., Liang, X., Liao, S., Lu, Z.: Pattern classification for ovarian tumors by integration of radiomics and deep learning features. Current Med. Imaging **18**(14), 1486–1502 (2022). https://doi.org/10.2174/1573405618666220516122145
13. Lof, P., et al.: Psychological impact of referral to an oncology hospital on patients with an ovarian mass. Int. J. Gynecologic Cancer **33**(1), January 2023. https://doi.org/10.1136/ijgc-2022-003753, https://ijgc.bmj.com/content/33/1/74, publisher: BMJ Specialist Journals Section: Original research
14. Ma, J., et al.: AbdomenCT-1K: is abdominal organ segmentation a solved problem? IEEE Trans. Pattern Anal. Mach. Intell. **44**(10), 6695–6714 (2022). https://doi.org/10.1109/TPAMI.2021.3100536. https://ieeexplore.ieee.org/document/9497733/
15. Meys, E.M.J., et al.: Subjective assessment versus ultrasound models to diagnose ovarian cancer: a systematic review and meta-analysis. Europ. J. Cancer **58**, 17–29 (2016). https://doi.org/10.1016/j.ejca.2016.01.007. https://www.sciencedirect.com/science/article/pii/S0959804916000459
16. Mulder, E.E., Gelderblom, M.E., Schoot, D., Vergeldt, T.F., Nijssen, D.L., Piek, J.M.: External validation of Risk of Malignancy Index compared to IOTA Simple Rules. Acta Radiologica (Stockholm, Sweden: 1987) **62**(5), 673–678 (2021). https://doi.org/10.1177/0284185120933990
17. Myronenko, A., Xu, Z., Yang, D., Roth, H., Xu, D.: Accounting for Dependencies in Deep Learning Based Multiple Instance Learning for Whole Slide Imaging (2021https://doi.org/10.48550/ARXIV.2111.01556, https://arxiv.org/abs/2111.01556, publisher: [object Object] Version Number: 1
18. Park, H., Qin, L., Guerra, P., Bay, C.P., Shinagare, A.B.: Decoding incidental ovarian lesions: use of texture analysis and machine learning for characterization and detection of malignancy. Abdominal Radiology **46**(6), 2376–2383 (2021). https://doi.org/10.1007/s00261-020-02668-3
19. Timmerman, D., et al.: ESGO/ISUOG/IOTA/ESGE Consensus Statement on preoperative diagnosis of ovarian tumors. Int. J. Gynecol. Cancer: Official J. Int. Gynecol. Cancer Soc. **31**(7), 961–982 (2021). https://doi.org/10.1136/ijgc-2021-002565
20. Togashi, K.: Ovarian cancer: the clinical role of US, CT, and MRI. Eur. Radiol. **13**(6), L87–L104 (2003) 10.1007/s00330-003-1964-y, https://doi.org/10.1007/s00330-003-1964-y
21. Woo, Y.L., Kyrgiou, M., Bryant, A., Everett, T., Dickinson, H.O.: Centralisation of services for gynaecological cancers - A Cochrane systematic review. Gynecol. Oncol. **126**(2), 286–290 (2012) 10.1016/j.ygyno.2012.04.012, https://www.sciencedirect.com/science/article/pii/S0090825812002673

FoTNet Enables Preoperative Differentiation of Malignant Brain Tumors with Deep Learning

Chenyi Hong[1], Hualiang Wang[2], Zhuoxuan Wu[3], Zuozhu Liu[1(✉)], and Junhui Lv[4(✉)]

[1] Zhejiang University - University of Illinois Urbana-Champaign Institute, Zhejiang University, Jiaxing, China
3180103573@zju.edu.cn, zuozhuliu@intl.zju.edu.cn
[2] The Hong Kong University of Science and Technology, Hongkong, China
hwangfd@ust.hk
[3] Department of Medical Oncology, Sir Run Run Shaw Hospital, College of Medicine, Zhejiang University, Hangzhou, China
wuzhuoxuan@zju.edu.cn
[4] Department of Neurosurgery, Sir Run Run Shaw Hospital, College of Medicine, Zhejiang University, Hangzhou, China
3415030@zju.edu.cn

Abstract. Glioblastoma (GBM), primary central nervous system lymphoma (PCNSL), and brain metastases (BM) are three common malignant central nervous system tumors. Accurate preoperative differentiation is essential for appropriate treatment planning and prognosis, however, it's challenging to differentiate these tumors using MRI due to their similar anatomical structures and imaging characteristics. In this paper, we first construct a new multi-center brain MRI dataset, including 315 training cases (GBM 64, PCNSL 59, BM 192) and 124 external test cases (24:23:77). Moreover, we propose a novel framework FoTNet for accurate diagnosis of the three tumors. Our model achieves a classification accuracy of 92.5% and an average AUC of 0.9754, outperforming previous methods. Our results demonstrates the great potential of AI in assisting physicians in differentiating between GBM, PCNSL, and BM, particularly in resource-limited clinical settings.

Keywords: brain tumor diagnosis · deep learning · multi-center dataset

1 Introduction

Glioblastoma (GBM), primary central nervous system lymphoma (PCNSL), and brain metastases (BM) are three common malignant neurosystem tumors [1,2]. In particular, GBM constitutes approximately 57% of all neurogliomas and 48% of primary malignant central nervous system (CNS) tumors [3–6]; around 20%

of cancers might lead to BM [2,7,8], predominantly originating from lung, breast etc.; PCNSL is a relatively rare tumor, accounting for about 3% to 4% of primary brain tumors and an incidence of 0.005‰ [9–12]. From the clinical perspective, accurately delineating these three tumors, especially in the preoperative stage with non-invasive differentiation methods, is of high necessity, as they demand significantly different treatment approaches and prognoses. For instance, BMs are typically managed through multimodal therapy, encompassing surgery, radiotherapy, chemotherapy, immunotherapy, and targeted therapy [13]; PCNSL treatment involves confirming its nature via lumbar puncture biopsy, followed by radiotherapy and chemotherapy [14]; GBMs are treated by striving for maximal safe resection, followed by concurrent radiochemotherapy and sequential chemotherapy [4].

Clearly distinguishing these three tumors remains a challenging task, while effective approaches or benchmark datasets are yet under development. First, the three tumors share lots of similar imaging characteristics, such as ring enhancement, significant peritumoral edema, watershed or gray-white matter interfaces, spherical shapes and multiple lesions etc. [18], imposing significant difficulty in neuroradiological differential diagnosis. Moreover, accurate diagnosis becomes more burdensome due to the uniform locational traits and non-distinctive lesion sizes. For instance, their radiological appearances might be very close to each other, with gray-white matter interfaces around the tumor area, internal necrotic zones, and a generally spherical shape, as displayed in Fig. 1. Last but not least, existing works for these tumors are still limited, especially for relatively rare PCNSL which might lead to data imbalance during learning. In consequence, constructing a larger and high-quality dataset for these three tumor plays a pivotal role towards clinically applicable decision-making for diagnosis.

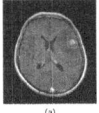 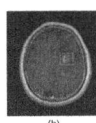 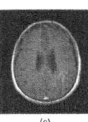

Fig. 1. MR images of (a) BM, (b) GBM, (c) PCNSL, see tumors in orange boxes. (Color figure online)

Previous attempts have been launched for accurate diagnosis of these tumors [15–17]. McAvoy et al. [15] utilizes the EfficientNetB4 to classify GBM and PCNSL, achieving an accuracy of 91–93%; Bathla et al. [16] proposes a network based on 3D ResNet to identify GBM, BM, and PCNSL, attaining a peak AUC of 0.854; Tariciotti et al. [17] employs ResNet101 for the classification of 64 GBM cases, 35 BM cases and 27 PCNSL cases, reaching a top AUC of 0.78. However, the accuracies of these works are not satisfactory as they usually employ general networks which are not tailored for these tumors. Meanwhile, these studies are trained and evaluated on relatively small-scale datasets, while their reliability and clinical applicability still need to be improved for practical applications.

In this paper, we propose a novel deep learning framework FoTNet for accurate classification of these three tumors. Specifically, we first construct a new cross-center dataset, comprising MRI data from public datasets UPENN and TCGA, as well as Sir Run Run Shaw Hospital affiliated with Zhejiang University School of Medicine, encompassing a total of 439 patients (88 GBM, 82 PCNSL, 289 BM). In each case, a team of professional researchers and clinicians meticulously extracted all 2D images containing tumor features. Moreover, we develop FoTNet, which employs a tailored ConvNeXt backbone to extract fine-grained features from MR images during training on the imbalanced dataset. We further incorporate patchify operations and the frequency channel attention (FCN) to prioritize the model's attention on regions relevant to tumor lesions and tumor contours. Our framework achieves a notable 92.50% classification accuracy on our test set, exceeding the state-of-the-art (SOTA) model [17] by 12.04% and outperforming other leading deep network models [19] by 2.52%. The integration of multi-center imaging data in the training and validation of our method substantially enhanced its robustness and reliability. The efficacy of our method has been thoroughly validated through extensive experiments.

2 Methods

2.1 Model Architecture

Basic Structure. Our model takes 2D MR brain images containing information on brain tumors as input. Initially, we preprocess input images through standardization to convert them into tensor inputs suitable for the network. Subsequently, the input tensor traverses multiple stages of convolutional networks before reaching the classification layer, where we determine the category with the highest probability. In addressing the unique characteristics and challenges of our classification task, we devise the following network structure (Fig. 2), drawing inspiration from the ConvNeXt architecture:

In the stem of our network, we employ a patchify operation using a 4×4 convolution with a stride of 4 to segment MR images into patches, enhancing the model's ability to focus on localized features and improve brain tumor classification while reducing computational load and parameter count. We incorporate a Frequency Channel Attention (FCA) Layer in each block to enhance the model's focus on tumor-relevant regions, leveraging different frequency components to enrich channel attention and improve classification accuracy. Our network utilizes a patch convolution layer instead of conventional downsampling methods to preserve spatial resolution and local feature extraction, reduces the use of activation and normalization layers to mitigate information loss, and replaces ReLU with GELU to improve feature learning for negatively responding pixels.

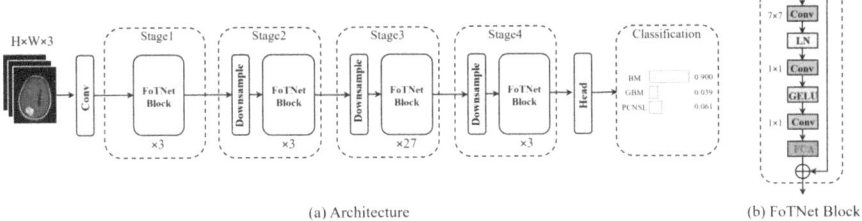

Fig. 2. The FoTNet. (a) Model structure. (b) FoTNet block with FCA layers.

Focus on Tumor Lesions. In our research, a key motivation is to guide the model's focus towards the tumor region within the MR image. In FoTNet, we employ the channel attention mechanism [20,21] which assigns different weight coefficients to different channels for focused learning on tumor lesions. Inspired by FcaNet [22], we utilize the FCA module which incorporates frequency components into the channel attention mechanism. By adopting this method, the model can focus on high-frequency components of MR images. The inclusion of high-frequency signals helps the model extract boundary information and detailed features of different tumors, such as shape and peritumoral edema etc., thereby enhancing tumor classification performance.

The two-dimensional (2D) DCT frequency spectrum $\mathcal{F} \in \mathbb{R}^{C \times H \times W}$ for the input feature map $\boldsymbol{X} \in \mathbb{R}^{C \times H \times W}$ is defined as:

$$\mathcal{F}_{i,h,w} = \sum_{j=0}^{H-1} \sum_{k=0}^{W-1} \boldsymbol{X}_{i,j,k} \boldsymbol{B}_{h,w}^{j,k}, \qquad (1)$$
$$\boldsymbol{B}_{h,w}^{j,k} = \cos(\frac{\pi h}{H}(j+\frac{1}{2})) \cos(\frac{\pi w}{W}(k+\frac{1}{2})),$$

where $\boldsymbol{B}$ indicates the basis function of 2D DCT [23]. Note that some constant normalization factors are removed for simplicity. In the FCA layer, the input $\boldsymbol{X}$ is first split into i parts along the channel dimensions, each part assigned a 2D DCT frequency component as follows:

$$\mathcal{F}_i = \mathrm{DCT}^{u_i, v_i}(\boldsymbol{X}_i) \quad = \sum_{j=0}^{H-1} \sum_{k=0}^{W-1} \boldsymbol{X}_{i,j,k} \boldsymbol{B}_{u_i,v_i}^{j,k}, \text{ s.t. } i \in \{0, 1, \cdots, n-1\}, \qquad (2)$$

in which $[u, v]$ are the frequency component 2D indices corresponding to $\boldsymbol{X}_i$, and $\mathcal{F}_i \in \mathbb{R}^{C'}$ is the C'-dimensional vector after the compression. The FCA module selects frequency components by first computing results for each component in channel attention, then selecting the top-k performing components based on these results. These selected components serve as the pre-processed inputs for channel attention. These components are then merged into a multi-spectral vector $\mathcal{F} = \mathrm{cat}([\mathcal{F}_0, \mathcal{F}_1, \cdots, \mathcal{F}_{n-1}])$. In this way, we integrate multiple frequency components, selecting the top 32 based on empirical evaluations, enhancing the traditional GAP-based channel attention which focuses only on the lowest frequency component [22]. This allows our FoTNet to better focus on tumor regions in MRI scans and to extract tumor features more precisely.

Following the incorporation of FCA, the output Y of the FoTNet block within our network can be calculated as: $Y = \psi(\mathcal{F})X + X_{input}$, where X_{input} represents the input tensor of the block, and $\psi()$ is a cascade module consisting of a fully connected layer and a sigmoid activation layer.

2.2 Model Training

The training process uses a single NVIDIA GeForce RTX 3090 GPU with PyTorch, a batch size of 1, the AdamW optimizer, and a constant learning rate of 3e-5, over 10 epochs, ensuring consistency across comparative experiments with identical parameters and pre-trained weights from ImageNet. We use focal loss as the loss function, with the parameter γ set to 2 and α_t set to 0.25.

3 Results

3.1 Dataset

We construct a new multi-center brain tumor MRI dataset. We collect MRI data of GBM, PCNSL, and BM from the Department of Neurosurgery at the Sir Run Run Shaw Hospital affiliated with Zhejiang University School of Medicine, with post-operative enhanced MR images of both typical and atypical cases. The dataset focuses on transverse T1-weighted contrast-enhanced MRI sequences. Experts select 2–18 tumor-containing slices from each patient's T1-weighted sequence. The GBM data includes 58 cases with 428 images; PCNSL has 82 cases with 343 images; BM contains 269 cases with 1,362 images. We further include external datasets: 18 cases with 617 2D images from UPENN-GBM and 12 cases with 81 2D images from TCGA-GBM. Ultimately, our dataset includes 88 GBM cases with 1,126 images, 82 PCNSL cases with 343 images, and 269 BM cases with 1,362 images. The dataset is split into training and test sets in a 5:2 ratio, i.e., 315 cases with 2,042 images for training, and 124 cases with 787 images for testing. All images are de-identified to ensure privacy.

3.2 Comparison with State-of-the-Art

The results of our FoTNet and baselines [24–32] are reported in Table 1. We report the classification accuracy and the average multi-class Area Under the Curve (AUC), where the 95% confidence intervals are also included with resampling. Our FoTNet achieves the highest classification accuracy and AUC scores of 92.50% and 0.9754 with the most compact confidence intervals, representing

Table 1. Test performance comparison on our new dataset.

model	model size	ACC/%	BACC/%	AUC (95%CI)
ImageNet-1K pre-trained models				
ResNet-50 [24]	25.56M	85.64 ± 0.52	60.44 ± 0.88	0.8548([0.813,0.894])
ResNet-101	44.55M	80.94 ± 0.76	49.95 ± 1.33	0.7959([0.753,0.839])
EfficientNet-B4 [25]	19.34M	82.41 ± 0.82	46.06 ± 1.67	0.8461([0.805,0.884])
EfficientNetV2-L [26]	118.52M	87.23 ± 0.69	65.54 ± 2.01	0.9052([0.871,0.935])
DenseNet-201 [27]	20.01M	83.54 ± 1.46	74.82 ± 0.72	0.9068([0.883,0.932])
VGG16_BN [28]	138.37M	82.96 ± 0.15	80.66 ± 0.65	0.9019([0.874,0.928])
VGG19_BN	143.68M	85.07 ± 0.83	81.85 ± 1.90	0.9086([0.880,0.937])
ResNeXt-50 [29]	25.03M	83.50 ± 0.49	47.68 ± 0.99	0.7963([0.761,0.832])
ResNeXt-101 [29]	88.79M	85.20 ± 0.82	67.53 ± 1.22	0.8905([0.860,0.918])
RegNetY-16G [30]	80.57M	86.97 ± 0.07	56.34 ± 0.29	0.8959([0.859,0.928])
ConvNeXt_large [19]	197.73M	85.77 ± 0.51	81.91 ± 1.33	0.9409([0.926,0.956])
ImageNet-22K pre-trained models				
ViT-B/16 [31]	86.42M	87.26 ± 0.67	73.33 ± 2.55	0.9190([0.898,0.944])
ViT-L/32	306.48M	84.32 ± 0.45	75.93 ± 0.75	0.8507([0.834,0.870])
MaxViT_xlarge [32]	474.24M	88.00 ± 0.38	79.45 ± 1.58	0.9515([0.936,0.964])
ConvNeXt_xlarge	348.04M	89.84 ± 0.14	77.29 ± 0.67	0.9640([0.953,0.974])
FoTNet(ours)	353.28M	**92.47± 0.03**	**87.51± 0.63**	**0.9754**([0.966,0.984])

a substantial improvement over baselines. This demonstrates that FoTNet is not only highly effective at differentiating among tumor images but also exhibits robustness against the inherent variability of tumors. We also calculate the balanced accuracy (BACC) to evaluate model performance, which averages the recall of each category with equal weight coefficients. Our model consistently outperforms baselines, especially in recognizing PCNSL, demonstrating better stability for classifying datasets with imbalance distributions.

Table 2. Ablations of the FCA module and focal loss (FL).

Methods	Accuracy	AUC
• FoTNet(without FCA and FL)	89.99%	0.9640
• FoTNet(without FCA)	90.75%	0.9674
• FoTNet(witout FL)	91.36%	0.9653
• FoTNet	**92.50%**	**0.9754**

3.3 Ablation Study

We investigate the effectiveness of the focal loss for imbalanced brain tumor classifications as well as the effectiveness of the FCA module in our network. The results are shown in Table 2. Upon the incorporation of the FCA layer, the model's performance improves significantly, with the accuracy rising from 89.99% to 91.36%. Employing focal loss further enhances the accuracy to 92.50%.

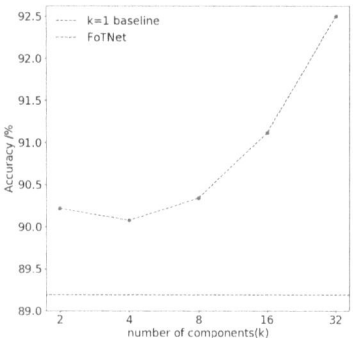

Fig. 3. Test set classification accuracy with top-k numbers of frequency components.

The Effects of Different Number of Frequency Components. We use the top-k highest-performing frequency components for the channel attention layer, with k being 1, 2, 4, 8, 16, or 32. As illustrated in the Fig. 3, our results show that using multiple frequency components improves performance compared to utilizing GAP channel attention, with the top-32 configuration achieving the best results, increasing classification accuracy from 90.22% to 92.50%.

3.4 Post-hoc Analysis

Confusion Matrix and PR Curves. The confusion matrix in Fig. 4(a) shows that our model has the highest true-positive rate for BM, indicating high sensitivity and specificity, but slightly lower accuracy for GBM and PCNSL. The model excels in recognizing BM but needs improvement in distinguishing between GBM and PCNSL.

We also present a comparative illustration of the PR (Precision-Recall) curves for the ConvNeXt and FoTNet models, which is shown in Fig. 4(b)(c). The BM PR curve remains stable with high precision, and the FCA module does not affect GBM diagnostic accuracy. Structural modifications significantly improve PCNSL precision at high recall, enhancing discrimination while preserving BM and GBM performance.

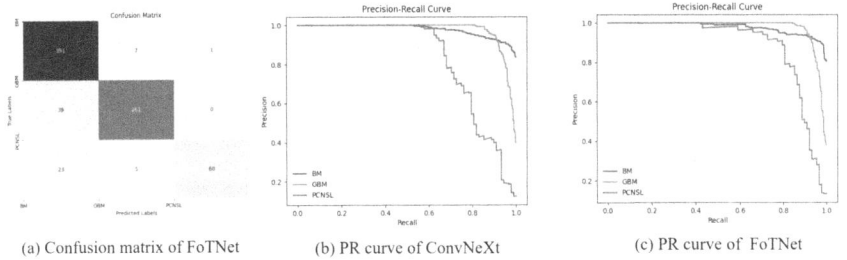

(a) Confusion matrix of FoTNet (b) PR curve of ConvNeXt (c) PR curve of FoTNet

Fig. 4. Confusion matrix and PR curves. (a) The confusion matrix of our best model. (b)(c) PR curve comparisons of classification models.

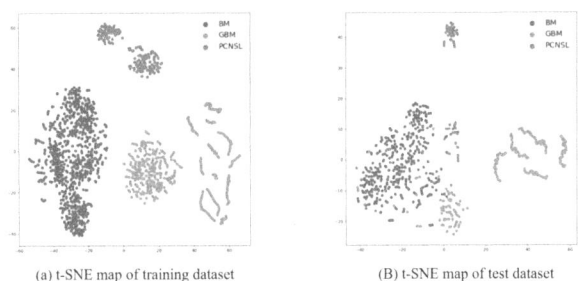

(a) t-SNE map of training dataset (B) t-SNE map of test dataset

Fig. 5. t-SNE visualization. Points of different colors represent various tumor samples.

T-SNE Visualization. We use the t-SNE algorithm to reduce dimensions and visualize model output vectors for the training and test datasets, as shown in Figure. 5. The training dataset visualization shows distinct tumor type clusters, indicating the model's ability to differentiate tumors, while the test dataset reveals some overlap, particularly between GBM and PCNSL. The t-SNE visualization highlights clear BM clusters with consistent features, whereas GBM and PCNSL clusters show greater variability, leading to some misclassifications.

Class Activation Map of Test Images. We employ the Smooth Grad-CAM++ method [33] to visualize the activation map as shown in Fig. 6. The figure highlights the model's precision in pinpointing the location of tumor lesions and making accurate classifications. However, in Fig. 6(b) (c), which are MR images of two GBM cases, the model not only focused on the tumor lesions but also on other high-signal areas in the brain. Such regions can also be observed in the brains of healthy individuals and do not contribute to the accurate classification of the tumor. Although the model correctly classifies the two presented GBM cases, such attention might lead to misjudgments in some tumor images.

Hard Cases Analysis. We analyze misclassified tumor images from the test dataset, shown in Fig. 7. Common issues include small lesions and interference

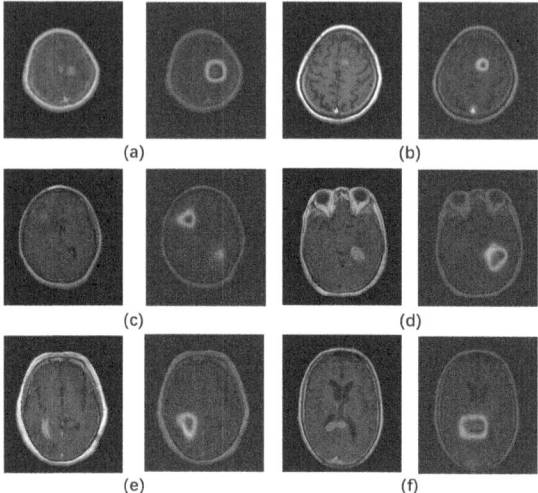

Fig. 6. CAM visualization. (a) A BM case, (b–c) GBM cases, (d–f) PCNSL cases.

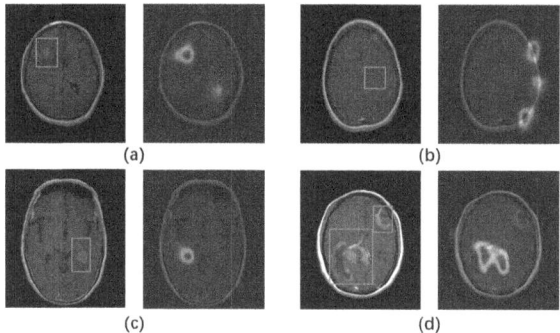

Fig. 7. Hard cases. (a–b) GBM cases, (c–d) PCNSL cases, see tumors in orange boxes. (Color figure online)

from normal brain structures like the choroid plexus, leading to the model's misfocus. For instance, in Fig. 7(d), the model misses features like small lesions and irregular shapes in PCNSL images. Improving the model's focus on tumor contours and distinguishing tumors from normal structures is crucial for better MRI classification.

4 Conclusion

In this paper, we construct a new dataset and introduce a novel deep learning frameworks FoTNet for accurate classification GBM, PCNSL and BM with MR images. Comprehensive experiments reveal the superiority of our model as compared to existing advanced CNN- and Transformer-based methods. Our work provides a promising solution for the preoperative classification of GBM, BM, and PCNSL, aiding clinicians in diagnoses and addressing healthcare access gaps when human and instrumental resources are limited.

Acknowledgements. This work is supported by the National Natural Science Foundation of China (Grant No. 62106222), the Natural Science Foundation of Zhejiang Province, China (Grant No. LZ23F020008), the Foundation of medical and health technology of Zhejiang province, China (2023RC189) and the Zhejiang University-Angelalign Inc. R&D Center for Intelligent Healthcare.

Disclosure of Interests. The authors have no relevant conflicts of interest to disclose.

References

1. Ostrom, Q.T., et al.: Cbtrus statistical report: primary brain and other central nervous system tumors diagnosed in the united states in 2012–2016. Neuro-oncology **21**(Supplement_5), v1–v100 (2019)
2. Nayak, L., Lee, E.Q., Wen, P.Y.: Epidemiology of brain metastases. Curr. Oncol. Rep. **14**, 48–54 (2012)
3. Ohgaki, H., Kleihues, P.: The definition of primary and secondary glioblastoma. Clin. Cancer Res. **19**(4), 764–772 (2013)
4. Tan, A.C., Ashley, D.M., López, G.Y., Malinzak, M., Friedman, H.S., Khasraw, M.: Management of glioblastoma: state of the art and future directions. CA: a cancer J. Clin. **70**(4), 299–312 (2020)
5. Ostrom, Q.T., et al.: Cbtrus statistical report: primary brain and central nervous system tumors diagnosed in the united states in 2008-2012. Neurooncology **17**(suppl_4), iv1–iv62 (2015)
6. Wirsching, H.G., Weller, M.: Glioblastoma. Malignant Brain Tumors: State-of-the-Art Treatment, pp. 265–288 (2017)
7. Tabouret, E., Chinot, O., Metellus, P., Tallet, A., Viens, P., Goncalves, A.: Recent trends in epidemiology of brain metastases: an overview. Anticancer Res. **32**(11), 4655–4662 (2012)
8. Barnholtz-Sloan, J.S., Sloan, A.E., Davis, F.G., Vigneau, F.D., Lai, P., Sawaya, R.E.: Incidence proportions of brain metastases in patients diagnosed (1973 to 2001) in the metropolitan detroit cancer surveillance system. J. Clin. Oncol. **22**(14), 2865–2872 (2004)
9. Jellinger, K., Radaskiewicz, T., Slowik, F.: Primary malignant lymphomas of the central nervous system in man. In: Malignant Lymphomas of the Nervous System: International Symposium, pp. 95–102. Springer (1975)
10. Commins, D.L.: Pathology of primary central nervous system lymphoma. Neurosurg. Focus **21**(5), 1–10 (2006)
11. Koeller, K.K., Smirniotopoulos, J.G., Jones, R.V.: Primary central nervous system lymphoma: radiologic-pathologic correlation. Radiographics **17**(6), 1497–1526 (1997)
12. Hochberg, F.H., Miller, D.C.: Primary central nervous system lymphoma. J. Neurosurg. **68**(6), 835–853 (1988)
13. Achrol, A.S., et al.: Brain metastases. Nature Reviews Disease Primers **5**(1), 5 (2019)

14. Batchelor, T., Loeffler, J.S.: Primary cns lymphoma. J. Clin. Oncol. **24**(8), 1281–1288 (2006)
15. McAvoy, M., et al.: Classification of glioblastoma versus primary central nervous system lymphoma using convolutional neural networks. Sci. Rep. **11**(1), 15219 (2021)
16. Bathla, G., et al.: Ai-based classification of three common malignant tumors in neuro-oncology: a multi-institutional comparison of machine learning and deep learning methods. J. Neuroradiol. (2023)
17. Tariciotti, L., Ferlito, D., Caccavella, V.M., Di Cristofori, A., Fiore, G., Remore, L.G., Giordano, M., Remoli, G., Bertani, G., Borsa, S., et al.: A deep learning model for preoperative differentiation of glioblastoma, brain metastasis, and primary central nervous system lymphoma: An external validation study. NeuroSci **4**(1), 18–30 (2022)
18. Usinskiene, J., et al.: Optimal differentiation of high-and low-grade glioma and metastasis: a meta-analysis of perfusion, diffusion, and spectroscopy metrics. Neuroradiology **58**, 339–350 (2016)
19. Liu, Z., Mao, H., Wu, C.Y., Feichtenhofer, C., Darrell, T., Xie, S.: A convnet for the 2020s. In: Proceedings of the IEEE/CVF Conference on Computer Vision and Pattern Recognition, pp. 11976–11986 (2022)
20. Hu, J., Shen, L., Sun, G.: Squeeze-and-excitation networks. In: Proceedings of the IEEE Conference on Computer Vision and Pattern Recognition, pp. 7132–7141 (2018)
21. Li, X., Wang, W., Hu, X., Yang, J.: Selective kernel networks. In: Proceedings of the IEEE/CVF Conference on Computer Vision and Pattern Recognition, pp. 510–519 (2019)
22. Qin, Z., Zhang, P., Wu, F., Li, X.: Fcanet: frequency channel attention networks. In: Proceedings of the IEEE/CVF International Conference on Computer Vision, pp. 783–792 (2021)
23. Ahmed, N., Natarajan, T., Rao, K.R.: Discrete cosine transform. IEEE Trans. Comput. **100**(1), 90–93 (1974)
24. He, K., Zhang, X., Ren, S., Sun, J.: Deep residual learning for image recognition. In: Proceedings of the IEEE Conference on Computer Vision and Pattern Recognition, pp. 770–778 (2016)
25. Tan, M., Le, Q.: Efficientnet: rethinking model scaling for convolutional neural networks. In: International Conference on Machine Learning, pp. 6105–6114. PMLR (2019)
26. Tan, M., Le, Q.: Efficientnetv2: smaller models and faster training. In: International Conference on Machine Learning, pp. 10096–10106. PMLR (2021)
27. Huang, G., Liu, Z., Van Der Maaten, L., Weinberger, K.Q.: Densely connected convolutional networks. In: Proceedings of the IEEE Conference on Computer Vision and Pattern Recognition, pp. 4700–4708 (2017)
28. Simonyan, K., Zisserman, A.: Very deep convolutional networks for large-scale image recognition. arXiv preprint arXiv:1409.1556 (2014)
29. Xie, S., Girshick, R., Dollár, P., Tu, Z., He, K.: Aggregated residual transformations for deep neural networks. In: Proceedings of the IEEE Conference on Computer Vision and Pattern Recognition, pp. 1492–1500 (2017)
30. Radosavovic, I., Kosaraju, R.P., Girshick, R., He, K., Dollár, P.: Designing network design spaces. In: Proceedings of the IEEE/CVF Conference on Computer Vision and Pattern Recognition, pp. 10428–10436 (2020)
31. Dosovitskiy, A., et al.: An image is worth 16x16 words: Transformers for image recognition at scale. arXiv preprint arXiv:2010.11929 (2020)

32. Tu, Z., Talebi, H., Zhang, H., Yang, F., Milanfar, P., Bovik, A., Li, Y.: Maxvit: Multi-axis vision transformer. In: European conference on computer vision. pp. 459–479. Springer (2022)
33. Omeiza, D., Speakman, S., Cintas, C., Weldermariam, K.: Smooth grad-cam++: an enhanced inference level visualization technique for deep convolutional neural network models. arXiv preprint arXiv:1908.01224 (2019)

Classification of Endoscopy and Video Capsule Images Using CNN-Transformer Model

Aliza Subedi[1]([✉]), Smriti Regmi[1], Nisha Regmi[2], Bhumi Bhusal[2], Ulas Bagci[2], and Debesh Jha[2]

[1] Pashchimanchal Campus, Pokhara, Nepal
alizasubedi4@gmail.com
[2] Machine and Hybrid Intelligence Lab, Department of Radiology, Northwestern University, Chicago, USA

Abstract. Gastrointestinal cancer is the leading cause of cancer-related incidence and death. Therefore, it is important to develop a novel computer-aided diagnosis system for early detection and enhanced treatment. Traditional approaches rely on the expertise of gastroenterologists to identify diseases. However, it is a subjective process, and the interpretation can vary even between expert clinicians. Considering recent progress in classifying gastrointestinal anomalies and landmarks in endoscopic and video capsule endoscopy images, this study proposes a hybrid model incorporating the advantages of Transformers and Convolutional Neural Networks (CNNs) for enhanced classification performance. Our model utilizes DenseNet201 as a CNN branch to extract local features and integrates the Swin Transformer branch for global feature understanding. Both of their features are combined to perform the classification task. For the GastroVision dataset, our proposed model demonstrates excellent performance with Precision, Recall, F1 score, Accuracy, and Matthews Correlation Coefficient (MCC) of 0.8320, 0.8386, 0.8324, 0.8386, and 0.8191, respectively, showcasing its robustness against class imbalance dataset and surpassing other CNNs as well as Swin Transformer model. Similarly, for the Kvasir-Capsule, a large video capsule endoscopy dataset, our model surpassed all other models, thereby achieving overall Precision, Recall, F1 score, Accuracy, and MCC of 0.7007, 0.7239, 0.6900, 0.7239, and 0.3871. Moreover, we generated saliency maps to explain our model's focus areas, showing its reliable decision-making process. The results underscore the potential of our hybrid CNN-Transformer model in aiding the early and accurate detection of gastrointestinal (GI) anomalies.

Keywords: Swin Transformer · Deep learning · Image classification · Gastrointestinal tract · GastroVision · Kvasir-Capsule

Supplementary Information The online version contains supplementary material available at https://doi.org/10.1007/978-3-031-73376-5_3.

1 Introduction

Medical image analysis has become increasingly essential in the diagnosis and prognosis of numerous medical conditions. One of the leading causes of cancer death is colorectal cancer [22]. It often begins as a growth called a polyp inside the colon or rectum. While not every polyp evolves into cancer, early identification and removal can halt the progression to cancer. Detecting these conditions early through screening significantly boosts survival chances, since it facilitates intervention during earlier, more treatable stages of the diseases. However, diagnosing these diseases is very time-consuming and tedious. Therefore, diagnostic tools enhanced with Artificial Intelligence (AI) have been developed lately to aid healthcare professionals in more effectively identifying and addressing these health issues. This approach also improves the quality of medical image analysis and reduces the time and resources needed for proper diagnosis, thereby improving the overall efficiency in healthcare sectors.

The advancement in deep learning algorithms, especially CNNs, which include blocks of convolutional layers, pooling layers, and fully connected layers, has significantly impacted computer vision tasks. Acting as foundational networks, CNNs have demonstrated remarkable effectiveness across various computer vision tasks, such as image classification [14], object detection [18], and image segmentation [4]. CNNs are inherently predisposed to identify patterns in data, providing flexible and adaptive operations that can capture the specific characteristics of the data. They are also effective when used to analyze images obtained from endoscopy [3], leading to more accurate and efficient diagnoses. Recent advancements in deep learning have witnessed the adaptation of transformers, originally designed for natural language processing tasks, to the domain of image analysis. A notable example of this transformative paradigm shift is the Swin Transformer, which demonstrates considerable potential for image classification tasks [15]. It offers a hierarchical vision by using shifted windows, enabling it to capture both global and local features of images.

This study presents an innovative approach to endoscopic image classification that combines the strengths of both CNNs and Swin Transformers. The proposed model, which uses a DenseNet201 [12] as the CNN branch and the Swin Transformer as the other branch, can efficiently classify endoscopic images. Our extensive experiments demonstrate the model's robustness and superior performance on the two medical GI datasets, paving the way for its application in clinical workflows.

The main contribution of our work is as follows:

1. We have presented a hybrid architecture for a gastrointestinal (GI) image classification task that effectively combines the capabilities of CNNs and Swin Transformer, bringing a new approach to GI image analysis.
2. We interpreted and visualized the performance of our models using saliency maps. This approach provides in-depth insights into the model's decision-making process about which parts of the input data are most influential in the model's predictions.

3. Our Hybrid model is able to achieve the MCC of 0.8191 for the GastroVision dataset [13] (highly imbalanced data with 22 classes) and MCC of 0.3871 for the Kvasir-Capsule dataset [24] (imbalance data with 11 classes), outperforming the standalone DenseNet201, Swin-T and other CNN methods across all evaluation metrics.

2 Related Work

Our work relates to CNN, Transformer, and GI tract diseases and findings. Here, we thoroughly review the literature that bears significant relevance to these areas.

2.1 Gastrointestinal Disease

Recent studies on the classification of endoscopic images have proposed several deep learning-based techniques [5,7,16]. For instance, Gamage et al. [9] presented the implementation of aggregation of deep learning features to predict anomalies related to digestive tract diseases. The ensemble involves the use of pre-trained CNN models such as DenseNet-201 [12], ResNet-18 [11], and VGG-16 [23] as the core feature extractors, integrated with a Global Average Pooling (GAP) layer. Similarly, Thambawita et al. [28] presented five different methods for GI tract disease classification task. Their combination of Densenet-161 and Resnet-152 with an additional MLP showcased highest performance. Afriyie et al. [2] presented Dn-CapsNets for identifying gastrointestinal tract diseases in the Kvasir-v2 dataset [19]. Srivastava et al. [25] introduced FocalConvNet, a network that merges focal modulation with lightweight convolutional layers, for the classification of anatomical and luminal findings within the Kvasir-Capsule dataset.

2.2 Transformer Network

Recently, architectures based on transformers have become popular due to their proficiency in handling long-term dependencies. The transformer architecture, which is commonly utilized for natural language processing, has been shown by Dosovitskiy et al. [8] to be effective when employed for computer vision tasks as well. According to their research, Vision Transformer (ViT) performs exceptionally well on image classification tasks when applied to sequences of image patches. Similarly, Touvron et al. [29] developed a novel teacher-student strategy for training convolution free transformers on ImageNet, using a distillation token to facilitate learning from the teacher through attention. Similarly, Usman et al. [30] and Matsoukas et al. [17] compared the performance of Transformers and CNN in various image classification tasks. Likewise, Tang et al. [27] proposed a Transformer-based Multi-task Network to analyze GI-tract lesions automatically by combining the benefits of both the CNN and Transformer.

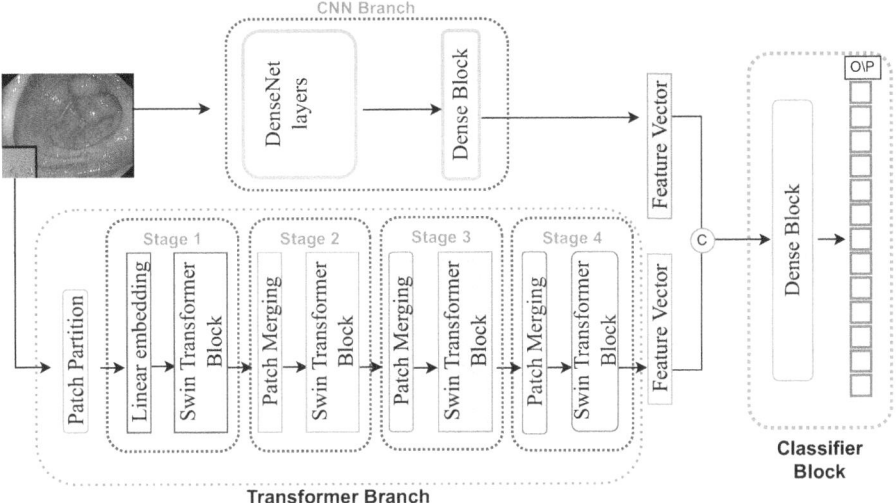

Fig. 1. Schematic representation of the Hybrid CNN-Transformer model.

3 Methodology

Here, we merged the capabilities of CNN and transformer networks to address the challenge of identifying the images encountered in endoscopic procedures. Our model has three components - *CNN branch*, *Transformer branch*, and *Classifier block* as shown in Fig. 1. The CNN branch captures detailed, local features while the Transformer branch focuses on broader, global aspects. These are subsequently merged into a single feature vector. This concatenated feature vector is fed into the *Classifier block* for the final classification task. The dense block in the figure consists of several layers, including a dense layer with 256 units, followed by a LeakyReLU activation with an alpha of 0.1, batch normalization, and a dropout layer with a 0.5 rate.

3.1 Transformer Branch

For the transformer branch, we utilized a variant of Swin Transformer, namely Swin-T, where 'T' refers to the tiny. The model was initialized with weights pre-trained on the ImageNet1K dataset [20]. Here, an input RGB image is first divided into non-overlapping patches of 4×4, each patch serving as a token. These patches have a feature dimension of $4 \times 4 \times 3 = 48$. Each patch is then linearly transformed into an embedding. The Swin Transformer model [15] consists of multiple layers of transformer encoders. Notably, the Swin Transformer replaces the traditional multi-head self-attention module in a Vision Transformer block with a shifting window-based self-attention. This design allows the model to attend to nearby patches while focusing on local context, without needing large attention windows. After processing the embeddings through several transformer layers, the information from various patches is aggregated to produce a single feature vector.

3.2 CNN Branch

In this branch, a CNN is used, specifically DenseNet201. We fine-tuned all the layers of the DenseNet201 on our dataset to enable it to extract features pertinent to our specific classification task. These densenet layers are then followed by *Dense block*.

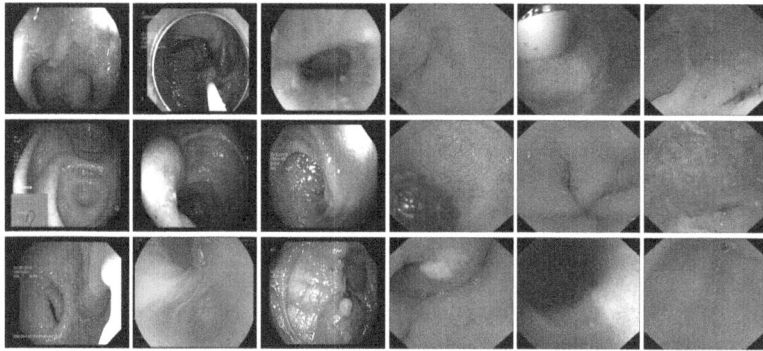

Fig. 2. Example samples from GastroVision [13] and Kvasir-Capsule [24]. The first three columns display images from GI endoscopy (GastroVision dataset), last three columns display from video capsule endoscopy (Kvasir-Capsule).

4 Dataset

We employed two multi-class publicly available GI endoscopy datasets: Gastro-Vision and the Kvasir-Capsule dataset. Some sample images of the datasets can be observed from the Fig. 2.

1. **GastroVision** [13]: The GastroVision dataset encompasses a wide variety of classes, including anatomical landmarks, pathological findings, cases of polyp removal, and normal or regular findings. A total of 7,930 images from 22 classes were used in the experiment following the dataset provider.
2. **Kvasir-Capsule** [24]: Kvasir-Capsule is the largest publicly available video capsule endoscopy dataset, which comprises 44,228 meticulously labeled images representing 13 distinct classes of anatomical and luminal findings. We have utilized 11 out of 13 classes, as some classes have very few samples.

5 Experiments

5.1 Implementation Details

Pre-training: In this study, all the models were implemented using Tensorflow [1] framework. We then resized all images to a standard dimension of 224 × 224 pixels and performed normalization and different data augmentation techniques to mitigate the data limitations. As part of the data augmentation process, we applied several techniques from TensorFlow's Keras utilities, including rescaling pixel intensities, inducing shear transformations, applying rotations of up to 30°, and performing vertical flips. Additionally, we adjusted the brightness to account for variations in lighting conditions. These strategic preprocessing and augmentation measures diversified our dataset and enhanced the generalization capabilities of our implemented architectures.

Experiment Setup and Configuration: Our experiments revolve around the task of classifying images using various architectures. To optimize our model, we used a trial-and-error approach to adjust various hyperparameters, such as learning rate, batch size, loss function, and optimizer. Specifically, the initial learning rate was set to $1e^{-2}$ for the GastroVision dataset and $1e^{-3}$ for the Kvasir-Capsule dataset, with dynamic adjustments made using the ReduceLROnPlateau scheduler to optimize convergence. We utilized categorical cross-entropy as the loss function, which is suitable for multi-class classification tasks. These hyperparameters were iteratively refined through multiple experimental runs to achieve the best performance metrics for each dataset. The GastroVision dataset is divided into an 80:20 ratio for training and testing, respectively. Similarly, we used official split (split1) for Kvasir-Capsule dataset.

5.2 Evaluation Metrics

We used various standard computer vision metrics to evaluate the performance of the models. These include MCC, weighted average precision, F1-score, recall, and overall accuracy. Of all these metrics, we prioritized MCC as the key metric since it is a more reliable statistical measure that yields a high score only when the classifier accurately predicts the majority of positive and negative data instances, and when both positive and negative predictions are mostly correct. We calculated these metrics over stratified 5-fold cross-validation and presented their averages along with their standard deviations. Additionally, we plotted saliency maps to visualize and interpret the performance of our model.

6 Results

We examined the trained model's effectiveness through multiple quantitative measures across a 5-fold cross-validation and the result is presented in Table 1 and Table 2. Our hybrid CNN-Transformer model's performance is compared

Table 1. Quantitative results on the GastroVision [13] dataset.

Method	Precision	Recall	F1-score	Accuracy	MCC
MobileNetV2 [21]	0.7308 ± 0.0356	0.7400 ± 0.0306	0.7318 ± 0.0332	0.7400 ± 0.0306	0.7083 ± 0.0347
ResNet50 [10]	0.7151 ± 0.0154	0.7320 ± 0.0159	0.7170 ± 0.0179	0.7320 ± 0.0159	0.6988 ± 0.0182
Xception [6]	0.7410 ± 0.0050	0.7499 ± 0.0032	0.7430 ± 0.0040	0.7499 ± 0.0032	0.7195 ± 0.0036
InceptionV3 [26]	0.7756 ± 0.0070	0.7847 ± 0.0049	0.7774 ± 0.0053	0.7860 ± 0.0048	0.7600 ± 0.0054
Densenet201 [12]	0.8056 ± 0.0062	0.8112 ± 0.0052	0.8046 ± 0.0056	0.8112 ± 0.0052	0.7886 ± 0.0059
Swin-T [15]	0.8075 ± 0.0023	0.8148 ± 0.0038	0.8082 ± 0.0031	0.8148 ± 0.0038	0.7924 ± 0.0042
DenseNet201 [12] + Swin-T [15]	**0.8320 ± 0.0204**	**0.8386 ± 0.0221**	**0.8324 ± 0.0250**	**0.8386 ± 0.0035**	**0.8191 ± 0.0038**

against the standalone DenseNet201, Swin Transformer model and other CNN methods. Here, each entry in the table is expressed as the mean ± SD of the respective performance metric, calculated over the 5-folds of cross-validation. We have reported the standard deviation (SD) to show the consistency of the model performance across five different folds, where a lower SD indicates consistent performance.

The classification performance of the models trained on GastroVision dataset is shown in Table 1. For this dataset, our proposed hybrid model, the combined Swin Transformer and DenseNet201, consistently outperformed the other models across all performance metrics. With a F1 score, accuracy, and MCC of 0.8324, 0.8386 and 0.8191 respectively, our model demonstrated the highest performance. Moreover, the table reveals that MCC of the hybrid model surpasses pretrained CNN methods by over 3% and exceeds Swin T by 2%. Most competitive to the Hybrid model is Swin T, with MCC of 0.7924. Moreover, the classification performance of various models trained on the Kvasir-Capsule dataset is presented in Table 2. The hybrid model achieved F1-score, Accuracy, and MCC of 0.6900, 0.7239, and 0.3871, respectively, outperforming all CNN-based methods and the standalone Swin Transformer model. Furthermore, as depicted in table, the MCC of the hybrid model trained on Kvasir-Capsule dataset outperforms pretrained CNN methods by more than 3% and surpasses Swin T by 2%. Swin T emerges as the closest competitor to the hybrid model. The disparity in the classification performance in two datasets can be attributed to the characteristics and quality of the datasets. The GastroVision dataset, with 22 classes, likely benefits from higher quality images and more distinct class features. In contrast, the Kvasir-Capsule dataset, despite its larger size, has significant class imbalance which poses a challenge for effective model training and performance.

7 Discussion

7.1 Limitations and Open Challenges

We encountered several problems, including data limitations and problems with resource usage during the study. Moreover, the inherent imbalance within the datasets added complexity to the training process, requiring careful handling to prevent biased model outcomes. We tried solving the problem of data limitations

Table 2. Quantitative results on the Kvasir-Capsule [24] dataset.

Method	Precision	Recall	F1-score	Accuracy	MCC
MobileNetV2 [21]	0.6100 ± 0.0047	0.6681 ± 0.0029	0.6283 ± 0.0028	0.6681 ± 0.0029	0.2489 ± 0.0074
ResNet50 [10]	0.6257 ± 0.0020	0.6810 ± 0.0026	0.6402 ± 0.0020	0.6810 ± 0.0026	0.2785 ± 0.0055
Xception [6]	0.5972 ± 0.0026	0.6857 ± 0.0012	0.6173 ± 0.0014	0.6857 ± 0.0012	0.2186 ± 0.0039
InceptionV3 [26]	0.5964 ± 0.0040	0.6861 ± 0.0019	0.6207 ± 0.0027	0.6861 ± 0.0019	0.2258 ± 0.0060
Densenet201 [12]	0.6726 ± 0.0041	0.7043 ± 0.0044	0.6711 ± 0.0035	0.7043 ± 0.0044	0.3508 ± 0.0067
Swin-T [15]	0.6951 ± 0.0081	0.7088 ± 0.0068	0.6785 ± 0.0082	0.7088 ± 0.0068	0.3600 ± 0.0173
DenseNet201 [12] +Swin-T [15]	**0.7007 ± 0.0238**	**0.7239 ± 0.0105**	**0.6900 ± 0.0180**	**0.7239 ± 0.0105**	**0.3871 ± 0.0333**

using various data augmentation techniques. To date, we have not yet engaged with clinicians to incorporate their invaluable expertise. Recognizing their critical role, we see a significant opportunity to collaborate with them to further refine our model's architecture and enhance its clinical relevance. Moving forward, our aim is to foster interdisciplinary collaboration, paving the way for the development of more effective and clinically applicable AI solutions.

7.2 Visualization Using Saliency Maps

A saliency map is a popular technique for visualizing the areas in the image that the model prioritized as the most important when making a prediction. To understand and interpret the behavior of our model, we have generated saliency maps for different representative classes, as shown in Fig. 3. These saliency maps are produced by computing the gradients of the model's output with respect to its input image, highlighting the pixels that have the most significant impact on the model's decision-making process.

In the figure, each class consists of four panels: the original image, the saliency map highlighting the regions with the highest influence on the model's decision, the overlay of the saliency map on the original image, and a binary mask derived from the saliency map. This comprehensive visualization helps in interpreting the model's focus areas, further strengthening its reliability and trustworthiness, which are essential in high-risk medical applications, such as GI disease classification tasks. For instance, in the *"Colon Polyps"* example, the saliency map emphasizes the area where the polyp is located, indicating that the model has learned to focus on the key feature relevant to diagnosing polyps. Similarly, for the *"Dyed Lifted Polyps"* class, the saliency map shows a strong focus on the dyed region, which is crucial for identifying the presence of a lifted polyp. These focused regions in the saliency maps correspond to the medically significant features that are critical for accurate classification, thereby demonstrating the model's interpretability. Additionally, in the *"Accessory Tools"* class, the saliency map highlights the regions where the medical tools are present in the endoscopic images. This level of interpretability is similarly applicable to other classes in the dataset, where the saliency maps consistently highlight the most relevant and significant features required for accurate classification.

By overlaying the saliency map on the original image, as depicted in Fig. 3, we can see a clear distinction between relevant and irrelevant areas. By provid-

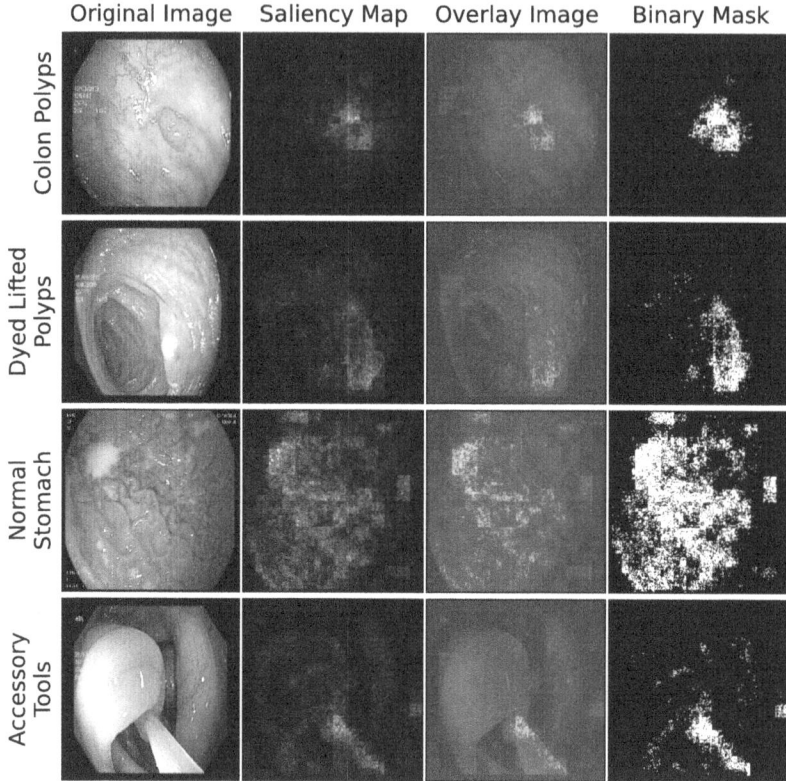

Fig. 3. Saliency maps visualization for GastroVision dataset for four representative classes. The Figure shows the significant regions in the input image that contribute to the model's decision.

ing a visual representation of what the model deems important, saliency maps reveal whether the model is focusing on appropriate features. This is particularly important in medical imaging, where the accuracy and reliability of the model's focus are critical for patient outcomes.

8 Conclusion

In this work, we proposed a hybrid model for GI image classification that combines the advantages of both CNN and Transformer for improved classification performance. The performance of our proposed models was evaluated using stratified 5-fold cross-validation and compared against individual DenseNet201, Swin Transformer model and other CNN methods. Our experimental results demonstrated that the proposed hybrid model consistently outperformed other methods, achieving an MCC of 0.8191 for the GastroVision dataset and an MCC of 0.3871 for the Kvasir-Capsule dataset. Furthermore, we employed saliency maps to visualize and interpret the decision-making process of our models. Additionally, the success of our model in classifying endoscopic images opens up new possibilities for its application in other medical imaging sectors, potentially enhancing clinical decision-making and improving patient well-being.

Acknowledgements. This project is supported by NIH funding: R01-CA246704, R01-CA240639, U01-DK127384-02S1, and U01-CA268808.

References

1. Abadi, M., et al.: Tensorflow: large-scale machine learning on heterogeneous distributed systems. arXiv preprint arXiv:1603.04467 (2016)
2. Afriyie, Y., A. Weyori, B., A. Opoku, A.: Gastrointestinal tract disease recognition based on denoising capsule network. Cogent Engineering **9**(1), 2142072 (2022)
3. Ahmed, A.: Classification of gastrointestinal images based on transfer learning and denoising convolutional neural networks. In: Proceedings of International Conference on Data Science and Applications: ICDSA 2021, vol. 1, pp. 631–639 (2022)
4. Alom, M.Z., Hasan, M., Yakopcic, C., Taha, T.M., Asari, V.K.: Recurrent residual convolutional neural network based on u-net (r2u-net) for medical image segmentation. arXiv preprint arXiv:1802.06955 (2018)
5. Chang, Y.Y., et al.: Deep learning-based endoscopic anatomy classification: an accelerated approach for data preparation and model validation. Surgical Endoscopy, pp. 1–11 (2021)
6. Chollet, F.: Xception: Deep learning with depthwise separable convolutions. In: Proceedings of the IEEE Conference on Computer Vision and Pattern Recognition, pp. 1251–1258 (2017)
7. Chou, C.K., Nguyen, H.T., Wang, Y.K., Chen, T.H., Wu, I.C., Huang, C.W., Wang, H.C.: Preparing well for esophageal endoscopic detection using a hybrid model and transfer learning. Cancers **15**(15), 3783 (2023)
8. Dosovitskiy, A., et al.: An image is worth 16×16 words: Transformers for image recognition at scale. arXiv preprint arXiv:2010.11929 (2020)
9. Gamage, C., Wijesinghe, I., Chitraranjan, C., Perera, I.: Gi-net: anomalies classification in gastrointestinal tract through endoscopic imagery with deep learning. In: 2019 Moratuwa Engineering Research Conference (MERCon), pp. 66–71 (2019)
10. He, K., Zhang, X., Ren, S., Sun, J.: Deep residual learning for image recognition. In: Proceedings of the IEEE Conference on Computer Vision and Pattern Recognition, pp. 770–778 (2016)
11. He, K., Zhang, X., Ren, S., Sun, J.: Identity mappings in deep residual networks. In: Leibe, B., Matas, J., Sebe, N., Welling, M. (eds.) ECCV 2016. LNCS, vol. 9908, pp. 630–645. Springer, Cham (2016). https://doi.org/10.1007/978-3-319-46493-0_38
12. Huang, G., Liu, Z., Van Der Maaten, L., Weinberger, K.Q.: Densely connected convolutional networks. In: Proceedings of the IEEE Conference on Computer Vision and Pattern Recognition, pp. 4700–4708 (2017)
13. Jha, D., et al.: Gastrovision: A multi-class endoscopy image dataset for computer aided gastrointestinal disease detection. arXiv preprint arXiv:2307.08140 (2023)
14. Li, Q., Cai, W., Wang, X., Zhou, Y., Feng, D.D., Chen, M.: Medical image classification with convolutional neural network. In: 2014 13th International Conference on Control Automation Robotics & Vision (ICARCV), pp. 844–848 (2014)

15. Liu, Z., et al.: Swin transformer: hierarchical vision transformer using shifted windows. In: Proceedings of the IEEE/CVF International Conference on Computer Vision (ICCV) (2021)
16. Lopez-Tiro, F., et al.: Boosting kidney stone identification in endoscopic images using two-step transfer learning. In: Mexican International Conference on Artificial Intelligence, pp. 131–141. Springer (2023). https://doi.org/10.1007/978-3-031-47640-2_11
17. Matsoukas, C., Haslum, J.F., Söderberg, M., Smith, K.: Is it time to replace cnns with transformers for medical images? arXiv preprint arXiv:2108.09038 (2021)
18. Papageorgiou, C.P., Oren, M., Poggio, T.: A general framework for object detection. In: Sixth International Conference on Computer Vision (IEEE Cat. No. 98CH36271), pp. 555–562 (1998)
19. Pogorelov, K., et al.: Kvasir: a multi-class image dataset for computer aided gastrointestinal disease detection. In: Proceedings of the 8th ACM on Multimedia Systems Conference, pp. 164–169 (2017)
20. Russakovsky, O., Deng, J., Su, H., Krause, J., Satheesh, S., Ma, S., Huang, Z., Karpathy, A., Khosla, A., Bernstein, M., et al.: Imagenet large scale visual recognition challenge. Int. J. Comput. Vision **115**, 211–252 (2015)
21. Sandler, M., Howard, A., Zhu, M., Zhmoginov, A., Chen, L.C.: Mobilenetv2: inverted residuals and linear bottlenecks. In: Proceedings of the IEEE Conference on Computer Vision and Pattern Recognition, pp. 4510–4520 (2018)
22. Siegel, R.L., Giaquinto, A.N., Jemal, A.: Cancer statistics, 2024. CA Cancer J. Clin. **74**(1), 12–49 (2024)
23. Simonyan, K., Zisserman, A.: Very deep convolutional networks for large-scale image recognition. arXiv preprint arXiv:1409.1556 (2014)
24. Smedsrud, P.H., et al.: Kvasir-capsule, a video capsule endoscopy dataset. Sci. Data **8**(1), 142 (2021)
25. Srivastava, A., Tomar, N.K., Bagci, U., Jha, D.: Video capsule endoscopy classification using focal modulation guided convolutional neural network. In: Proceedings of the IEEE 35th International Symposium on Computer-Based Medical Systems (CBMS), pp. 323–328 (2022)
26. Szegedy, C., Vanhoucke, V., Ioffe, S., Shlens, J., Wojna, Z.: Rethinking the inception architecture for computer vision. In: Proceedings of the IEEE Conference on Computer Vision and Pattern Recognition, pp. 2818–2826 (2016)
27. Tang, S., Yu, X., Cheang, C.F., Liang, Y., Zhao, P., Yu, H.H., Choi, I.C.: Transformer-based multi-task learning for classification and segmentation of gastrointestinal tract endoscopic images. Comput. Biol. Med. **157**, 106723 (2023)
28. Thambawita, V., Jha, D., Riegler, M., Halvorsen, P., Hammer, H.L., Johansen, H.D., Johansen, D.: The medico-task 2018: Disease detection in the gastrointestinal tract using global features and deep learning. In: Proceedigns of the Medico 2018 (2018)
29. Touvron, H., Cord, M., Douze, M., Massa, F., Sablayrolles, A., Jégou, H.: Training data-efficient image transformers & distillation through attention. In: Proceedings of the International Conference on Machine Learning, pp. 10347–10357 (2021)
30. Usman, M., Zia, T., Tariq, A.: Analyzing transfer learning of vision transformers for interpreting chest radiography. J. Digit. Imaging **35**(6), 1445–1462 (2022)

Multimodal Deep Learning-Based Prediction of Immune Checkpoint Inhibitor Efficacy in Brain Metastases

Tobias R. Bodenmann[1,2], Nelson Gil[1], Felix J. Dorfner[1], Mason C. Cleveland[1], Jay B. Patel[1], Shreyas Bhat Brahmavar[1], Melisa S. Guelen[1], Dagoberto Pulido-Arias[1], Jayashree Kalpathy-Cramer[1], Jean-Philippe Thiran[2], Bruce R. Rosen[1], Elizabeth Gerstner[1], Albert E. Kim[1(✉)], and Christopher P. Bridge[1(✉)]

[1] Athinoula A. Martinos Center for Biomedical Imaging, Massachusetts General Hospital and Harvard Medical School, Charlestown, MA, USA
akim46@mgh.harvard.edu, cbridge@mgh.harvard.edu
[2] Ecole Polytechnique Fédérale de Lausanne, Lausanne, Switzerland

Abstract. Recent studies demonstrate promising efficacy with immune checkpoint inhibitors (ICI) for brain metastases (BM), an unmet need in modern oncology. However, a predictive biomarker for ICI efficacy is needed to inform precision-based use of ICI given its high toxicity rate. Here, we present several multimodal deep learning (DL) approaches that integrate pre-treatment magnetic resonance imaging (MRI) and clinical metadata to predict ICI efficacy for BM. Using a multi-institutional dataset of 548 patients, our best-performing models achieve an AUROC of 0.674 (±0.041). In future work, we will accrue additional clinical and radiologic data to improve performance. Furthermore, our work thus far will serve as a baseline by which to trial alternate fusion strategies to improve and refine multimodal biomarker discovery for precision oncology.

Keywords: brain metastases · checkpoint inhibitor efficacy · deep learning · multimodal integration

1 Introduction

Brain metastases (BM) are a significant challenge in modern oncology due to increasing incidence and limited treatments. To this end, we and other groups have demonstrated that immune checkpoint inhibitors (ICI), which augment T-cell cytotoxicity against tumor cells, hold promise for treatment-refractory BM

A.E. Kim and C.P. Bridge—Contributed equally.

Supplementary Information The online version contains supplementary material available at https://doi.org/10.1007/978-3-031-73376-5_4.

of diverse tumor types [3,8,22]. However, this promising efficacy of ICI must be carefully balanced with the risk of toxicity, as more than half of patients treated with ICI experience at least one grade-3 or higher adverse event [3,8,22]. Despite functional imaging and comprehensive genomic analysis, a robust predictive biomarker for ICI response across different primary tumor types has not been identified [14,15]. To date, most commercial biomarkers are limited to one modality, which reflects only a limited facet of tumor biology without the broader clinical or biological context.

Using a deep learning (DL) classification pipeline, we present an exploratory application of multimodal fusion of patient-matched radiologic and clinical data to predict ICI efficacy for BM. As recent efforts have linked quantitative size and texture analysis of tumors on conventional radiology with CD8 T-cell infiltration and ICI efficacy for multiple tumor types [6,16,21,23], we hypothesized the different sequences within multiparametric brain magnetic resonance imaging (MRI) would reflect complementary facets of tumor biology and contain signal for ICI efficacy prediction. In addition, as different modalities provide complementary biological data, we hypothesized that a multimodal model would have improved predictive performance beyond that of any individual modality. To the best of our knowledge, our effort is one of the first to predict ICI efficacy and will add to the growing body of work for multimodal fusion applied to cancer biomarker discovery.

2 Methodology

2.1 Dataset

In collaboration with the Massachusetts General Hospital (MGH), Dana-Farber Cancer Institute (DFCI), and Brigham and Women's Hospital (BWH), a comprehensive multi-institutional dataset was established prior to this study. This dataset comprises paired radiologic (MRI) and clinical data for a total of 548 patients with brain metastases (BM) who received immune checkpoint inhibitors (ICI), including 197 patients from MGH, 194 from BWH, and 157 from DFCI. The MRI dataset consisted of the following sequences, if available: T1 pre-contrast (T1pre), T1 post-contrast (T1post), T2, fluid-attenuated inversion recovery (FLAIR), susceptibility-weighted imaging (SWI), and apparent diffusion coefficient (ADC). Clinical data are also presented in Tab. S1 and included the following features: patient characteristics (age, gender, body mass index (BMI)), treatment information (type of ICI administered, use of steroids), primary tumor histology, laboratory values (hemoglobin, albumin levels, platelet count, absolute neutrophil and lymphocyte count), and radiological findings (anatomical location and number of BM present in the pre-treatment scan, BM volume).

Due to the limited dataset, we employed 10-fold cross-validation (CV) and split on the patient-level into 80% for training, 10% for validation, and 10% for testing in each fold. Average Area under the Receiver Operating Curve (AUROC) performance and its standard deviation (std) are presented in Sect. 3.

2.2 Image Preprocessing

We employed a standard preprocessing pipeline for our MRI images. The pipeline included resampling the scans to 1mm isotropic resolution, registering all scans within a study to the corresponding T1post scan [12], and applying N4 bias field correction [2] and skull stripping [13]. We further performed longitudinal registration between the pre- and post-treatment scans and normalized the image intensities to zero mean and unit variance.

2.3 Generation of Ground Truth Labels

We trained models using two separate ground truth labels. One model predicted ICI on the individual tumor level (e.g., metastasis-level model). For this model, intracranial response, on the level of each individual BM, was used as the ground truth label. Often patients with BM harbor multiple BM, which will display mixed response across spatially separated BM in the same patient (e.g., one BM will respond but another may progress) [3,8]. The pre-treatment MRI was the latest MRI before the start of ICI treatment. The post-treatment MRI was the MRI obtained at 6 months (or latest available scan if treatment duration was less than 6 months) while on ICI treatment. Using the T1post sequence, an in-house fine-tuned BM segmentation model was used to segment and track BM in both scans [20]. By quantifying volumetric changes between the pre- and post-treatment scan, each BM was assigned a response classification based on the Response Assessment in Neuro-Oncology (RANO) criteria [17]. To maximize class balance and demonstrate clinical utility, we converted the four classes within RANO (e.g., complete response (CR), partial response (PR), stable disease (SD), progressive disease (PD)) into a binary classification problem by combining CR, PR, SD into a single class (e.g., response) and using PD as the other class (e.g., progression). Using the five largest BM for each patient, response labels were obtained for each BM.

We also trained a model that predicted response on the patient-level. To obtain a patient-level response label, we summed the total volume of the five largest pre-treatment BM and their respective counterparts in the post-treatment scan. Using the volumetric changes across the two timepoints, we assigned each patient the corresponding label (e.g., response vs progression).

2.4 Clinical Model

To address missing clinical information, we used a median imputer for numerical features and one-hot-encoded categorical features. The data was split into training (80%) and testing (20%) sets separately for the random forest classifier, which was trained using 5-fold cross-validation and evaluated on the test set.

We conducted a grid search to explore 360 combinations of hyperparameters, including the number of decision trees, maximum tree depth, and minimum samples for splitting. We then filtered out non-important features to address the issue of neural networks performing poorly on tabular data [9,10]. The eight most predictive features were used as input to a neural network, which was trained using a median imputer, standard scaling, and quintile transformation. We then used a simple feedforward network with ReLU activation functions for three hidden layers of 8, 16, and 32 neurons respectively, and sigmoid activation for the output layer. We employed an Adam optimizer, a reduce-on-plateau learning rate scheduler, and Bayesian optimization using Optuna [1] to optimize the hyperparameters, including the learning rate, dropout probability, and number of neurons in the hidden layers.

2.5 MRI Model

Preprocessing: Given the varying size of BM, we used a fixed $40 \times 40 \times 40$ mm region-of-interest (ROI) around each BM, based on the tumor segmentation. This ROI encompassed the entire contrast-enhancing tumor and adjacent peritumor region. Additionally, we explored a masking approach. For this strategy, we reapplied the segmentation mask to the ROI, dilated this mask by a small offset to allow for potential imperfect segmentations, and zeroed out everything outside this dilated mask. This approach prevented the model from overfitting to spurious features, while preserving critical information about BM size and texture.

Data Augmentation: To maximize the diversity of our training images and enhance model generalizability, we employed data augmentation techniques. We applied affine transformations, flips, rotations, Gaussian noise and smoothing, and introduced intensity shifts and contrast adjustments. Additionally, we simulated missing MRI sequences by randomly omitting and replacing them with black images, enhancing the model's robustness to real-world scenarios.

Hyperparameters: We explored a variety of hyperparameters, including learning rates, schedulers, 3D versus 2D approaches, pretrained and non-pretrained networks, varying architectures from common ResNets and DenseNets to simple custom CNNs, weight decays, label smoothing, number of epochs, and other parameters. Capitalizing on recent breakthroughs in foundation models, we further explored linear probing on top of the vision encoder from the Radiology Foundation Model (RadFM) developed by Wu et al. [24].

Model Architecture: Initial experiments revealed that standard ResNets and DenseNets led to excessive overfitting and focused on irrelevant image features unrelated to the BM. To address this, we explored simpler architectures by designing a custom CNN with varying layer and neuron counts. Our results showed that a 3-layer configuration struck a balance between model simplicity and sufficient complexity to extract meaningful information from the input patches, and therefore yielded the best performance.

For the metastasis-level model, input patches from the $40 \times 40 \times 40$ mm ROI for the 6 MRI sequences were fed into the simple 3-layer CNN. For the patient-level model, all BM belonging to the same patient were fed through the first 2 convolutional layers, with weights shared across the different BM. A reduction operation was then applied, taking the median of the resulting feature maps across the BM similar to the approach of Havaei et al. [11]. The resulting feature map was then fed through the final convolutional layer.

2.6 Multimodal Model

To combine the two unimodal models previously described, we investigated three different fusion strategies, as illustrated in Fig. 1.

Intermediate Fusion: We used a simple 3-layer CNN to extract features from the BM ROI, which was comprised of six different sequences. In parallel, we processed the 8 most predictive clinical features through a fully connected layer to obtain a feature representation of the clinical data. Subsequently, we concatenated the MRI feature representation vector with the clinical feature representation vector, which was then fed into another fully connected layer with a sigmoid activation function, ultimately yielding the prediction of ICI efficacy.

Co-attention Guided Fusion: This strategy leverages information from one modality to inform feature extraction from another modality. This approach recognizes that specific clinical data (e.g., primary tumor histology, ICI type) may correspond to distinct radiologic phenotypes [18]. After extraction of clinical and imaging features, the co-attention mechanism computed an attention score between each clinical feature and each MRI feature, resulting in an attention map. This map highlighted the most relevant MRI features for each clinical feature. Finally, the weighted sum of these values, guided by the attention map, was used to create a refined MRI representation that integrates the clinical context. This enriched representation was then fused with the original clinical features for the downstream prediction task.

Late Fusion: For this approach, we obtained separate predictions using a 3-layer CNN for the MRI data and 3 fully connected layers for the clinical data. The final prediction was then derived by averaging the individual predictions.

3 Results and Discussion

3.1 Clinical Model

Using a Random Forest model, we identified clinical features that were most predictive of treatment response (see also Fig. S1). On the held-out test set, the top features included pre-treatment BM volume, laboratory values (e.g., platelets, hemoglobin), and patient characteristics (e.g., BMI and age). The location of the BM, tumor histology, and type of ICI administered displayed relatively low predictive power for ICI efficacy. When quantifying single feature contributions for ICI efficacy prediction, BM volume emerged as the most crucial factor. This is consistent with prior work for prediction of response to radiation for BM [4,7]. In addition, we found that the levels of platelets, albumin, and hemoglobin had moderate correlation to ICI response. These blood markers are linked to systemic and anti-tumor inflammation, which has emerged in prior work as potential predictors of ICI response [5]. In addition, BMI and age were predictive of treatment

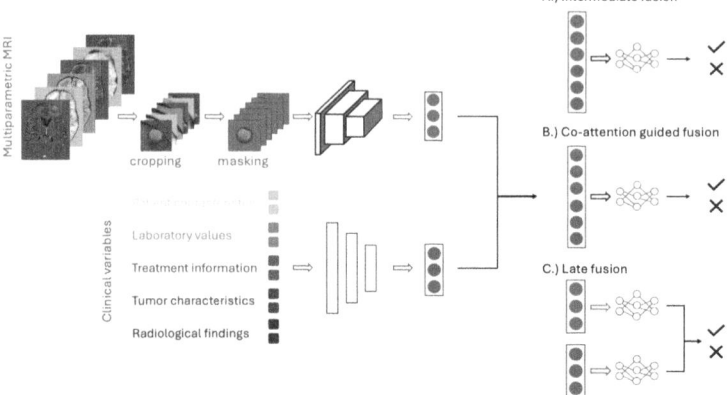

Fig. 1. Schematic overview of the multimodal integration pipeline. The clinical and MRI features are combined in three ways: A.) concatenation of both modalities, B.) use of clinical feature to guide the MRI feature extraction, C.) training of separate models for each modality.

response. This finding is consistent with medical intuition that a patient's pre-treatment functional status is often linked to treatment outcomes.

The full random forest model, integrating all clinical metadata, achieved an AUROC of 0.651 (±0.044) on the test set. Notably, a random forest model trained with only the 8 most predictive features, resulted in an AUROC of 0.643 (±0.042), demonstrating effective feature selection without compromising performance. A neural network using the top 8 clinical features achieved an AUROC of 0.618 (±0.038), setting a baseline for clinical model performance. Overall, the analysis is consistent with the growing body of work suggesting that clinical data possess predictive power, albeit weak, for ICI efficacy [5].

Table 1. Average AUROC performances (±std) evaluated using 10-fold CV comparing the clinical and MRI-only baseline models.

Data	Model type	Implementation details	AUROC	
			Validation	Test
Clinical	NN	8 features, patient-level	0.631(±0.035)	0.618(±0.038)
MRI	custom CNN	T1post, BM-level (cropped-only)	0.599(±0.038)	0.601(±0.053)
MRI	custom CNN	T1post, BM-level	0.676(±0.046)	0.656(±0.058)
MRI	RadFM	T1post, BM-level	0.669(±0.056)	0.636(±0.048)
MRI	custom CNN	T1post, patient-level	0.681(±0.042)	0.640(±0.040)
MRI	custom CNN	all six sequences, BM-level	0.672(±0.046)	0.646(±0.046)
MRI	custom CNN	all six sequences, patient-level	0.676(±0.045)	0.638(±0.042)

3.2 MRI Model

First, we determined the optimal ROI size, encompassing the tumor and peritumor area, as input to the model. We trialed different ROI sizes ranging from $30 \times 30 \times 30$ mm to $50 \times 50 \times 50$ mm. Larger ROIs would predominantly contain non-tumor areas and may increase the risk of spurious correlations for an exploratory task. We ultimately achieved optimal performance with a $40 \times 40 \times 40$ mm ROI centered around the BM as input, with an AUROC of 0.601 (± 0.053) on the test set. Interestingly, we found that our masking approach, using the $40 \times 40 \times 40$ mm ROI, demonstrated a significantly higher AUROC of 0.656 (± 0.058) (Table 1). This observation may be due to the masking approach 'forcing' our model to focus on the BM, rather than non-tumor areas.

Metastasis-Level Model: When determining response on the metastasis-level, our MRI-based DL model demonstrated a test set performance of AUROC of 0.656. While this performance was higher than that of the clinical model (Table 1), our MRI-only model achieved modest predictive performance for ICI efficacy. Additionally, using the RadFM foundation model as a feature extractor, our model did not demonstrate significant improvement (AUROC of 0.636). As the biological significance of imaging patterns may vary with histology, we will train separate histology-specific models and compare these models to the overall model encompassing all tumor types. In future work, we will accrue more MRI scans from melanoma and non-small cell lung cancer BM patients, as these tumor types are the most likely to benefit from ICI [8, 22]. Histology-specific models may demonstrate improved predictive performance.

Multiparametric Model: As different sequences included in MRI reflect complementary facets of tumor biology, we hypothesized that a model integrating all MRI sequences would demonstrate improved performance compared to a T1post-only model. Given studies that link degree of blood extravasation [19], which is reflected in the SWI sequence, with immune activation, our results are surprising. We anticipate that these results may be in part due to our small and heterogenous training set. As stated above, our future work will be to re-train our models with a larger and more homogeneous (e.g., histology-specific) dataset.

Patient-Level Model: When determining response on the patient-level, our MRI-based DL model failed to yield improved predictive performance. Our patient-level model obtained an AUROC of 0.640 (vs 0.656 for the metastasis-level model) for a T1post-only model and 0.638 (vs 0.646) for a model integrating all six sequences (Table 1). We believe that our relatively small dataset may be a primary explanation for this outcome, as capturing the complex interplay between multiple metastases within a patient can be challenging with a small and heterogeneous dataset.

3.3 Multimodal Fusion Model

The effect of integrating clinical metadata to medical imaging in augmenting predictive performance for biomarker development remains under-explored in

Table 2. Average AUROC performances (±std) evaluated using 10-fold CV comparing different fusion mechanisms of the multimodal fusion model.

Model type	Implementation details	AUROC	
		Validation	Test
Multimodal	T1post, intermediate fusion	0.677(±0.045)	0.658(±0.045)
Multimodal	T1post, co-attention guided fusion	0.667(±0.043)	0.648(±0.058)
Multimodal	T1post, late fusion	0.689(±0.039)	0.674(±0.041)

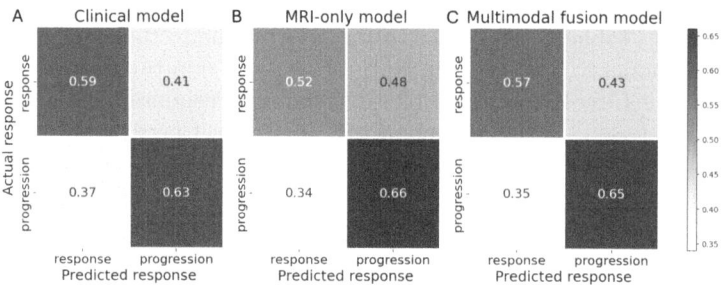

Fig. 2. *Confusion matrices for A) the clinical model, B) the MRI-only model and C) the multimodal fusion model.*

solid tumor oncology. To this end, we explored fusing clinical metadata with brain MRI to predict ICI efficacy using multiple strategies (Table 2). The late fusion approach served as our baseline experiment. Our late fusion multimodal model resulted in a similar performance to our MRI-based model. The best-performing MRI-based model demonstrated an AUROC of 0.656 (±0.058) on the test set, whereas the late fusion model had an AUROC of 0.674 (±0.041). Our intermediate fusion multimodal model, which demonstrated an AUROC of 0.658 (±0.045) on the test set, also did not yield significant improvement over the MRI-based model. Similarly, with guided fusion, using clinical metadata to guide MRI feature extraction did not enhance prediction of ICI efficacy (AUROC 0.648 (±0.058)). Taken together, our experiments suggest that multimodal fusion may not always result in improved performance. Furthermore, the lack of large multimodal datasets is a major challenge for evaluating different multimodal machine learning strategies in oncology, highlighting a need for multi-institutional collaboration and distributed learning methods. To overcome these limitations, our future efforts will incorporate additional data, as well as patient-matched histo-pathology and genomic data as inputs to our multimodal model.

Finally, an analysis of the confusion matrices in Fig. 2 reveals a consistent trend across all models, characterized by a higher proportion of false negatives, where true responders are misclassified as non-responders. This translates to higher specificity (approximately 65%) compared to sensitivity (approximately 57%). For this application, prioritizing high specificity is crucial, as this ensures that patients receiving treatment are truly likely to benefit, thereby minimizing unnecessary treatment toxicity for non-responding patients.

4 Conclusion

Our study presents one of the first efforts to integrate complementary data modalities to develop a predictive biomarker for ICI response, an unmet need in solid tumor oncology. We found that a MRI-based end-to-end DL model demonstrated modest signal for the prediction of ICI efficacy. Perhaps unexpectedly, the integration of clinical variables did not significantly enhance performance. These results will serve as a baseline by which to measure performance for other approaches using additional data or alternate fusion strategies.

Given the relatively underexplored nature of multimodal biomarker development in clinical oncology, we believe our data thus far may be of interest. These results motivate multi-institutional cohort assembly to improve and refine multimodal biomarker discovery. As a next step, we will accrue additional data from specific cohorts (e.g., patients with melanoma, NSCLC BM) to train histology-specific models. We will also trial other feature extraction frameworks from other radiology foundation models to improve classification performance. Finally, we may explore ordinal classification and measures to boost model interpretability. Importantly, we hope our work motivates future multimodal DL studies to improve patient stratification and advance precision oncology.

References

1. Akiba, T., Sano, S., Yanase, T., Ohta, T., Koyama, M.: Optuna: a next-generation hyperparameter optimization framework. In: Proceedings of the 25th ACM SIGKDD International Conference on Knowledge Discovery and Data Mining (2019)
2. Beare, R., Lowekamp, B., Yaniv, Z.: Image segmentation, registration and characterization in r with SimpleITK **86**, 8. https://doi.org/10.18637/jss.v086.i08
3. Brastianos, P.K., et al.: Pembrolizumab in brain metastases of diverse histologies: phase 2 trial results **29**(7), 1728–1737. https://doi.org/10.1038/s41591-023-02392-7
4. Cho, S.J., et al.: Prediction of treatment response after stereotactic radiosurgery of brain metastasis using deep learning and radiomics on longitudinal mri data . https://doi.org/10.1038/s41598-024-60781-5
5. Chowell, D., et al.: Improved prediction of immune checkpoint blockade efficacy across multiple cancer types
6. Dercle, L., et al.: Early readout on overall survival of patients with melanoma treated with immunotherapy using a novel imaging analysis **8**(3), 385–392. https://doi.org/10.1001/jamaoncol.2021.6818
7. DeVries, C., et al.: Prediction of brain metastasis response to stereotactic radiosurgery using mri and machine learning: effects of primary cancer site and metastasis volume **114**(3S)

8. Goldberg, S.B., et al.: Pembrolizumab for management of patients with NSCLC and brain metastases: long-term results and biomarker analysis from a non-randomised, open-label, phase 2 trial **21**(5), 655–663. https://doi.org/10.1016/S1470-2045(20)30111-X
9. Gorishniy, Y., Rubachev, I., Khrulkov, V., Babenko, A.: Revisiting deep learning models for tabular data. http://arxiv.org/abs/2106.11959
10. Grinsztajn, L., Oyallon, E., Varoquaux, G.: Why do tree-based models still outperform deep learning on tabular data? http://arxiv.org/abs/2207.08815
11. Havaei, M., Guizard, N., Chapados, N., Bengio, Y.: HeMIS: etero-modal image segmentation. https://doi.org/10.48550/ARXIV.1607.05194. https://arxiv.org/abs/1607.05194, version Number: 1
12. Hoffmann, M., Billot, B., Greve, D.N., Iglesias, J.E., Fischl, B., Dalca, A.V.: SynthMorph: learning contrast-invariant registration without acquired images. https://doi.org/10.1109/TMI.2021.3116879. http://arxiv.org/abs/2004.10282
13. Hoopes, A., Mora, J.S., Dalca, A.V., Fischl, B., Hoffmann, M.: Synthstrip: skull-stripping for any brain image. NeuroImage **260**, 119474 (2022). https://doi.org/10.1016/j.neuroimage.2022.119474. https://www.sciencedirect.com/science/article/pii/S1053811922005900
14. Jenkins, R.W., Barbie, D.A., Flaherty, K.T.: Mechanisms of resistance to immune checkpoint inhibitors **118**(1), 9–16. https://doi.org/10.1038/bjc.2017.434. https://www.nature.com/articles/bjc2017434
15. Kim, A.E., et al.: Abnormal vascular structure and function within brain metastases is linked to pembrolizumab resistance **26**(5), 965–974. https://doi.org/10.1093/neuonc/noad236, https://academic.oup.com/neuro-oncology/article/26/5/965/7467048
16. Ligero, M., et al.: A CT-based radiomics signature is associated with response to immune checkpoint inhibitors in advanced solid tumors **299**(1), 109–119. https://doi.org/10.1148/radiol.2021200928
17. Lin, N.U., et al.: Response Assessment in Neuro-Oncology (RANO) group: Response assessment criteria for brain metastases: proposal from the RANO group **16**(6), e270–278. https://doi.org/10.1016/S1470-2045(15)70057-4
18. Lipkova, J., et al.: Artificial intelligence for multimodal data integration in oncology **40**(10), 1095–1110. https://doi.org/10.1016/j.ccell.2022.09.012. https://linkinghub.elsevier.com/retrieve/pii/S153561082200441X
19. Nourshargh, S., Alon, R.: Leukocyte migration into inflamed tissues **41**(5), 694–707. https://doi.org/10.1016/j.immuni.2014.10.008. https://linkinghub.elsevier.com/retrieve/pii/S1074761314003847
20. Patel, J., et al.: A deep learning based framework for joint image registration and segmentation of brain metastases on magnetic resonance imaging. In: Deshpande, K., Fiterau, M., Joshi, S., Lipton, Z., Ranganath, R., Urteaga, I., Yeung, S. (eds.) Proceedings of the 8th Machine Learning for Healthcare Conference. Proceedings of Machine Learning Research, vol. 219, pp. 565–587. PMLR (11–12 Aug 2023), https://proceedings.mlr.press/v219/patel23a.html
21. Sun, R., et al.: A radiomics approach to assess tumour-infiltrating CD8 cells and response to anti-PD-1 or anti-PD-l1 immunotherapy: an imaging biomarker, retrospective multicohort study **19**(9), 1180–1191. https://doi.org/10.1016/S1470-2045(18)30413-3
22. Tawbi, H.A., et al.: Combined nivolumab and ipilimumab in melanoma metastatic to the brain **379**(8), 722–730. https://doi.org/10.1056/NEJMoa1805453. http://www.nejm.org/doi/10.1056/NEJMoa1805453

23. Trebeschi, S., et al.: Prognostic value of deep learning-mediated treatment monitoring in lung cancer patients receiving immunotherapy **11**, 609054. https://doi.org/10.3389/fonc.2021.609054
24. Wu, C., Zhang, X., Zhang, Y., Wang, Y., Xie, W.: Towards generalist foundation model for radiology by leveraging web-scale 2d&3d medical data. http://arxiv.org/abs/2308.02463

Seeing More with Less: Meta-learning and Diffusion Models for Tumor Characterization in Low-Data Settings

Eva Pachetti[1,2]([✉]) and Sara Colantonio[1]

[1] Institute of Information Science and Technologies, National Research Council, Pisa, Italy
eva.pachetti@isti.cnr.it
[2] Department of Information Engineering, University of Pisa, Pisa, Italy

Abstract. While deep learning excels in many areas, its application in medicine is hindered by limited data, which restricts model generalizability. Few-shot learning has emerged as a potential solution to this problem. In this work, we leverage the strengths of meta-learning, the primary framework for few-shot learning, along with diffusion-based generative models to enhance few-shot learning capabilities. We propose a novel method that jointly trains a diffusion model and a feature extractor in an episodic-based manner. The diffusion model learns conditional generation based on each episode's support samples. After updating its parameters, it generates additional support samples for each class. The augmented support set is used to train a feature extractor within a prototypical meta-learning framework. Notably, we propose a weighted prototype computation based on the distance between each generated sample and the original class prototype, i.e., derived solely from the original support samples. Evaluations on two tumor characterization tasks (prostate cancer aggressiveness and breast cancer malignity assessment) demonstrate our approach's effectiveness in improving prototype representation and boosting classification performance. Find our code at: https://github.com/evapachetti/meta_diffusion.

Keywords: Few-shot learning · Meta-learning · Diffusion models

1 Introduction

Deep learning (DL) models in medical imaging hold significant promise for clinical applications. However, insufficient data makes the training phase challenging, often resulting in inaccurate and unreliable models for real-world use. A promising solution to this data scarcity problem is few-shot learning (FSL) [5,6], which aims to enable practical DL model training in data-limited scenarios, such as medical imaging [2,13,28]. One of the most popular frameworks for addressing FSL is meta-learning. Meta-learning involves two main objectives: an inner objective related to a specific classification task (or *episode*) and an outer objective where the model learns to distinguish data more generally. This approach, also known as *episodic* training, enhances the model's ability to learn robust features, improving generalization, an essential property in data-scarce domains.

Within meta-learning, a popular technique is the prototype-based approach [19]. In each episode, the idea is to calculate a prototype for each class based on the few training examples available in that episode (known as the *support* set). New images within the episode (*query* samples) are then classified by measuring their distance to each class prototype. Various techniques have been proposed to enhance prototype informativeness. Cao et al. [1] leveraged the similarity between classes to calibrate prototypes of new classes using base classes learned beforehand. He et al. [9] introduced a transformer-based module to extract more informative prototypes. Zhang et al. [26] tackled the problem of low-informative prototypes utilizing prior knowledge and pre-trained features obtained from complete prototypes of well-represented classes.

Another strategy employed in literature to enhance prototype informativeness involves using generative models. Zhang et al. [27] proposed a prototype meta-hallucination approach to generate more informative prototypes by hallucinating additional support samples. Specifically, they meta-trained a Variational Autoencoder (VAE) [16] to learn the distribution of inter-sample differences and synthesized newly labeled samples by fusing the sampled inter-sample difference and each given support sample. However, VAEs often struggle with capturing the full complexity of images due to their reliance on a Gaussian-distributed latent space, leading to blurry or imprecise image generation [3,12] and, thus, to less meaningful prototypes.

In recent years, the emergence of diffusion-based models has resulted in generative models exhibiting unprecedented capabilities [15,17]. Indeed, diffusion models overcame the limitations provided by generative models such as VAEs and Generative Adversarial Networks (GANs) [7], including the need for estimation of intractable the normalizing constant of the probability function, the need for network constraints, and the training instability, leading to the generation of more realistic and informative samples that closely resemble actual data.

In this work, we leverage the power of Denoising Diffusion Probabilistic Models (DDPMs) [10,20] to enhance prototype informativeness in a prototypical meta-learning framework for few-shot medical image classification. Specifically, we propose a training method integrating real and synthetic data within a joint episodic training process between the DDPM and a feature extractor. Support samples are provided to train a conditional DDPM during each episode, which then hallucinates additional data samples for each class. These synthetic samples supplement the limited actual data available for prototype construction. Furthermore, we implement a dynamic weighting mechanism to ensure the synthetic samples enhance rather than confuse the model. This method calculates weights for the synthetic samples based on their proximity to the original data-derived prototype. By integrating actual and synthetic data with dynamic weighting, we aim to forge more reliable prototypes, thereby enhancing classification performance in low-data scenarios.

2 Methods

2.1 Proposal

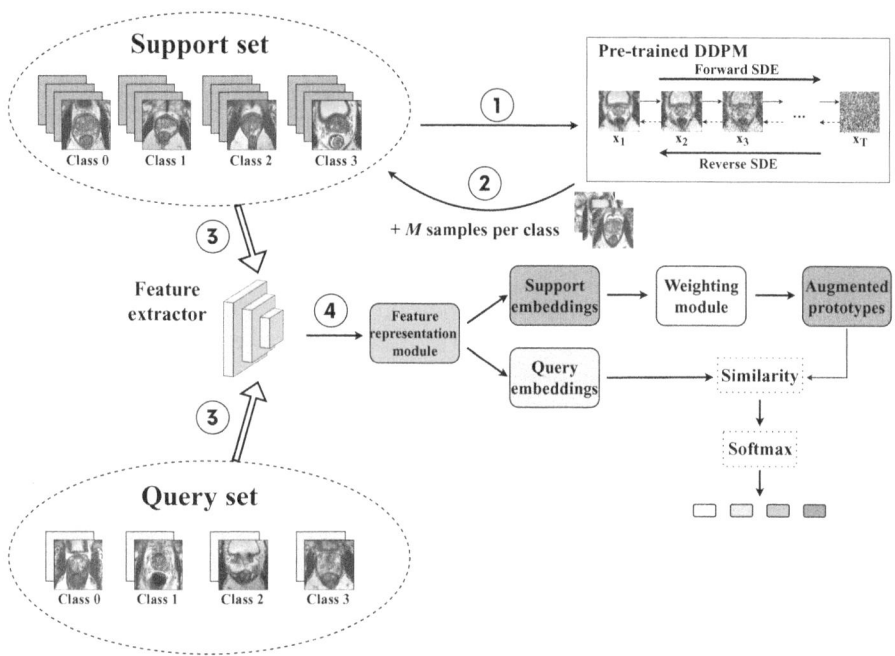

Fig. 1. Illustration of the proposed approach for a single episode. Each episode consists of a support set and a query set. All support samples are provided to the DDPM for conditional training, guided by the class embedding. The DDPM parameters are updated to generate M additional support samples for each class. The augmented support and query sets are fed through a feature extractor to compute feature maps. These maps are then processed by a dedicated feature representation module, whose function depends on the chosen prototypical framework (e.g., average pooling for ProtoNet). The weighting module (Fig. 2) adjusts the augmented support embeddings based on their similarity to the original prototype. The original and generated support embeddings, weighted accordingly, are averaged to compute the augmented prototype for each class. A distance calculation between query embeddings and augmented prototypes follows. Finally, classification is achieved using a softmax distribution over these computed similarities.

We provide a detailed overview of our proposed strategy in Fig. 1. As a first step, to provide the DDPM with prior knowledge, we pre-train it on an unlabeled dataset to perform unconditioned generation. This step aims to equip the model with an understanding of the invariant features in the anatomical structures regardless of the specific class. Following this pre-training step, we perform a joint episodic training between the DDPM and a convolutional feature extractor. Within each episode, consisting of a support set (containing N images per class) and a query set, we utilize the entire support set as input to the pre-trained DDPM to train it on performing conditioned generation leveraging the class embedding, as proposed by Ho et al. [11]. After updating its parameters, we employ the DDPM to generate M additional support samples for each class. These synthetic samples are added to the support set and used to train the feature extractor.

Especially during the initial training phase, the DDPM's synthetic samples may not match the quality and detail of the real support images. This discrepancy can result in less accurate prototypes. To address this challenge, we propose a weighting module (see Fig. 2) that implements a dynamic weighting approach when constructing prototypes. We define $S_c = \{(\boldsymbol{x}_1, y_1), \ldots, (\boldsymbol{x}_N, y_N)\}$ as the support set of class c, containing only the original support samples, and $\widetilde{S}_c = \{(\tilde{\boldsymbol{x}}_1, y_1), \ldots, (\tilde{\boldsymbol{x}}_M, y_M)\}$ the support set containing only generated samples of that class. Given a feature extractor f_ϕ, we calculate the real prototype, i.e., built on the original support samples only as follows:

$$\boldsymbol{p}_c = \frac{\sum_{i=1}^{|S_c|} f_\phi(\boldsymbol{x}_i)}{|S_c|}. \qquad (1)$$

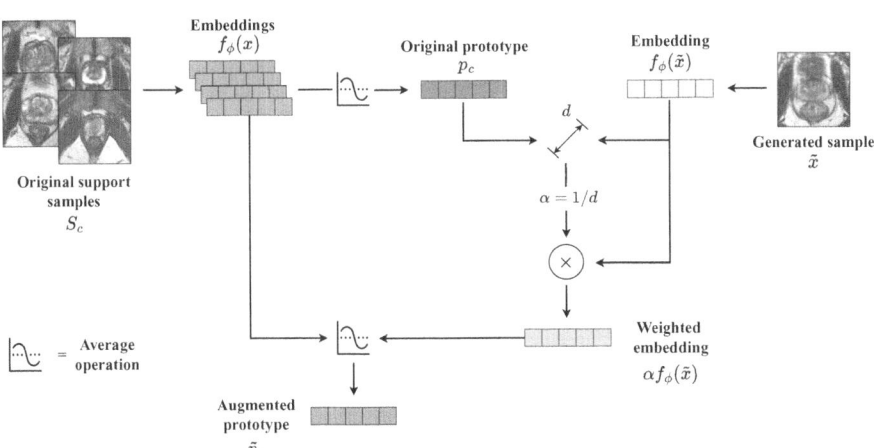

Fig. 2. Representation of the weighting module for augmented prototype estimation. For clarity, we depicted the weighting module functioning for a single class and a single generated sample.

The weight for a generated support sample $\tilde{\boldsymbol{x}}_j^c$ of class c is provided by calculating the reciprocal of the Euclidean distance between the generated support sample embedding $f_\phi(\tilde{\boldsymbol{x}}_j^c)$ and the class prototype computed on the original support samples $\boldsymbol{p}_c$:

$$\alpha_j^c = \frac{1}{d(f_\phi(\tilde{\boldsymbol{x}}_j^c), \boldsymbol{p}_c))}. \tag{2}$$

Finally, the prototype for the class c is calculated as follows:

$$\tilde{\boldsymbol{p}}_c = \frac{\sum_{i=1}^{|S_c|} f_\phi(\boldsymbol{x}_i) + \sum_{j=1}^{|\tilde{S}_c|} \alpha_j^c f_\phi(\tilde{\boldsymbol{x}}_j)}{|S_c| + \sum_{j=1}^{|\tilde{S}_c|} \alpha_j^c}. \tag{3}$$

After calculating the final prototypes for each class, we measure the Euclidean distance between each query sample and the corresponding prototypes. Finally, we apply the softmax function to these distances to predict the class labels for the query samples.

2.2 Model Details

Generative Model. Traditional DDPMs corrupt data through a finite number of noise steps and train a sequence of probabilistic models to reverse each step of this noise corruption. Unlike the classical approach, we adopt a continuous diffusion approach proposed by Song et al. [21]. This method employs a continuous noise-perturbation process modeled by stochastic differential equations (SDEs). A reverse-time SDE describes the inverse diffusion process from noise to image, and samples can be generated by solving this equation using numerical SDE solvers [21]. In our experiments, we described the diffusion process using a sub-variance-preserving SDE [21] and employed the Euler-Maruyama numerical solver [14] for image generation.

Prototypical Framework. We evaluated our approach leveraging three well-known prototypical meta-learning frameworks: Prototypical Network (ProtoNet) [19], Meta Deep Browninan Distance Covariance (Meta DeepBDC) [24] and Covariance Network (CovNet) [23]. These methods all share the common principle of constructing a class prototype by averaging the embeddings of the support samples belonging to that class. Classification is then performed by measuring the distance between the embedding of each query sample and all class prototypes. However, they diverge in their approach to feature representation: ProtoNet utilizes a basic first-order representation via average pooling, CovNet leverages a second-order representation with the covariance matrix, and Meta DeepBDC goes beyond pairwise relationships by considering the joint distribution between features through the computation of the BDC matrix.

2.3 Experiments

We evaluated our approach on two tumor characterization tasks: prostate cancer aggressiveness (PI-CAI dataset [18]) and breast cancer lesion classification

(BreakHis dataset [22]). For prostate cancer, we classified tumors based on four (2–5) prognostic scores defined by the International Society of Urological Pathology (ISUP) [4]. We pre-trained our DDPM on all available benign lesions (11202 images) from PI-CAI dataset. Then, we performed episodic supervised learning using all cancerous lesions (1611 training, 200 validation, 238 testing from a total of 2049). In breast cancer classification, we used images from various magnifications (40X, 100X, 200X; 6090 images total) for DDPM unconditioned pre-training and focused on 400X magnification images (1475 training, 165 validation, 183 testing from a total of 1819) for episodic training, performing binary classification (benign vs. malignant).

We evaluated our approach using k-shot classification tasks ($k \in \{1, 2, 3, 4, 5\}$) for each prototypical framework. We investigated the effect of generating one, two or three synthetic support samples per class to improve prototype informativeness. The training process involved 100 epochs, each with 50 meta-training and 50 meta-validation episodes. We assessed performance using mean and standard deviation (STD) of Area Under the ROC Curve (AUROC) across 50 meta-testing episodes. We employed a ResNet-18 for feature extraction, with a learning rate of 10^{-4} and weight decay of 10^{-2}, trained using AUC margin loss [25] and the Proximal Epoch Stochastic optimizer [8]. For the DDPM, we utilized a learning rate of 20^{-4} and 400000 steps for training, with 1000 noise scales (variance from 0.1 to 20) during image generation. The training phase was conducted using the negative log-likelihood loss function and the Adam optimizer.

3 Results and Discussion

Our experimental results (Table 1 and Table 2) demonstrate that generating synthetic support samples improves performance (AUROC) across all three prototypical methods for both classification tasks. On the prostate cancer aggressiveness classification task, ProtoNet benefits the most, achieving an AUROC increase of over 11% (from 0.634 to 0.749) in the *5-shot* setting with two synthetic samples per class. On the other hand, for breast lesion classification, CovNet exhibits the most considerable absolute improvement (15%, from 0.514 to 0.664) in the *4-shot* setting with three synthetic samples, even though ProtoNet achieves the highest AUROC (0.785) with three synthetic samples in the *5-shot* setting.

Figure 3 visually explores the relationship between the number of synthetic support samples and AUROC performance. Interestingly, the results indicate that performance gains are not always monotonic. This implies that adding more synthetic samples may not always yield the best results. For example, the CovNet model on the BreakHis dataset achieves its peak AUROC with a single synthetic support sample in the *1-shot* setting. Adding more synthetic samples, in this case, actually leads to decreased performance. This behavior may be due to low-quality image generation (e.g., images lacking features of the desired class), which the weighting method failed to mitigate. However, this phenomenon

Table 1. Mean and STD (in brackets) of AUROC across 50 meta-test episodes on the PI-CAI dataset (prostate cancer aggressivenes classification).

Framework	K-shot	Baseline	+1 support	+2 support	+3 support
ProtoNet	1-shot	0.527 (0.067)	0.578 (0.087)	0.564 (0.065)	**0.628 (0.094)**
	2-shot	0.570 (0.080)	0.599 (0.088)	0.597 (0.072)	**0.631 (0.066)**
	3-shot	0.590 (0.074)	0.595 (0.133)	0.610 (0.104)	**0.634 (0.067)**
	4-shot	0.609 (0.067)	0.594 (0.084)	0.677 (0.088)	**0.698 (0.083)**
	5-shot	0.634 (0.095)	0.678 (0.091)	**0.749 (0.089)**	0.616 (0.087)
Meta DeepBDC	1-shot	0.579 (0.061)	0.580 (0.058)	**0.660 (0.124)**	0.658 (0.084)
	2-shot	0.600 (0.062)	0.598 (0.049)	0.563 (0.094)	**0.660 (0.046)**
	3-shot	0.595 (0.068)	0.627 (0.070)	0.637 (0.082)	**0.646 (0.090)**
	4-shot	0.612 (0.044)	0.538 (0.093)	0.656 (0.065)	**0.664 (0.077)**
	5-shot	0.632 (0.099)	0.646 (0.076)	0.662 (0.039)	**0.667 (0.073)**
CovNet	1-shot	0.503 (0.041)	0.527 (0.057)	0.552 (0.064)	**0.570 (0.048)**
	2-shot	0.529 (0.045)	0.558 (0.052)	0.560 (0.080)	**0.585 (0.059)**
	3-shot	0.575 (0.075)	0.588 (0.043)	**0.652 (0.074)**	0.619 (0.065)
	4-shot	0.580 (0.032)	0.599 (0.081)	0.619 (0.087)	**0.623 (0.068)**
	5-shot	0.602 (0.053)	0.610 (0.044)	0.626 (0.083)	**0.631 (0.080)**

Table 2. Mean and STD (in brackets) of AUROC across 50 meta-test episodes on the BreakHis dataset (breast lesion classification).

Framework	K-shot	Baseline	+1 support	+2 support	+3 support
ProtoNet	1-shot	0.588 (0.197)	0.572 (0.163)	0.618 (0.212)	**0.655 (0.205)**
	2-shot	0.651 (0.186)	0.674 (0.252)	0.678 (0.160)	**0.762 (0.151)**
	3-shot	0.644 (0.176)	0.679 (0.210)	**0.718 (0.163)**	0.698 (0.178)
	4-shot	0.665 (0.163)	0.679 (0.199)	0.740 (0.165)	**0.762 (0.140)**
	5-shot	0.724 (0.151)	0.732 (0.145)	0.778 (0.159)	**0.785 (0.158)**
Meta DeepBDC	1-shot	0.524 (0.194)	0.514 (0.212)	**0.544 (0.208)**	0.539 (0.216)
	2-shot	0.524 (0.204)	0.536 (0.213)	0.562 (0.198)	**0.576 (0.198)**
	3-shot	0.589 (0.187)	0.591 (0.179)	0.609 (0.196)	**0.633 (0.170)**
	4-shot	0.534 (0.181)	0.537 (0.205)	0.575 (0.203)	**0.632 (0.178)**
	5-shot	0.594 (0.181)	0.625 (0.176)	0.578 (0.190)	**0.666 (0.180)**
CovNet	1-shot	0.506 (0.143)	**0.536 (0.131)**	0.495 (0.226)	0.534 (0.199)
	2-shot	0.516 (0.103)	0.560 (0.179)	0.590 (0.187)	**0.618 (0.191)**
	3-shot	0.512 (0.062)	0.513 (0.083)	0.579 (0.140)	**0.643 (0.224)**
	4-shot	0.514 (0.154)	0.578 (0.175)	0.611 (0.158)	**0.664 (0.175)**
	5-shot	0.532 (0.103)	0.549 (0.084)	0.589 (0.168)	**0.642 (0.187)**

is also evident in baseline configurations (without adding synthetic data), where increasing the number of support samples (e.g., from *1-shot* to *2-shot*) does not always lead to improved performance. For example, in Table 2, the Meta DeepBDC model demonstrates a decrease in AUROC of approximately 5% when transitioning from a *3-shot* to a *4-shot* setting. This could be due to redundant information or poor-quality samples.

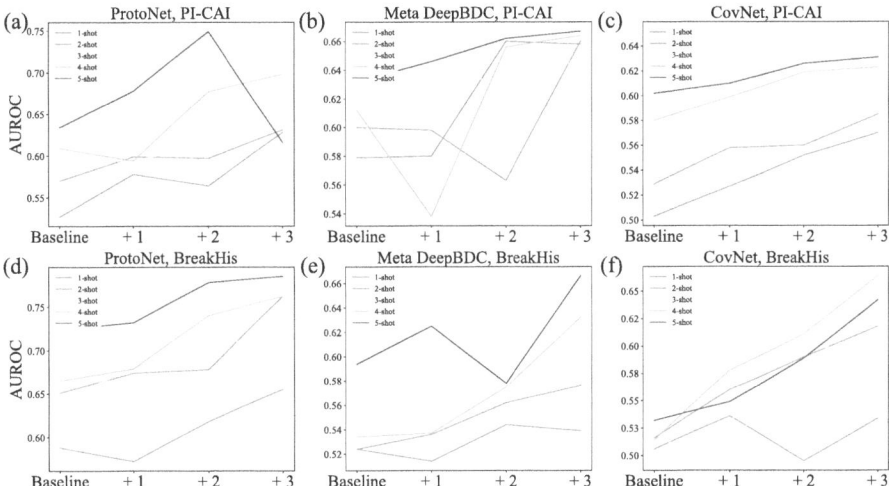

Fig. 3. Illustration of classification performance in terms of AUROC as the number of synthetic support samples per class increases. Best viewed in color. (a) ProtoNet with PI-CAI; (b) Meta DeepBDC with PI-CAI; (c) CovNet with PI-CAI; (d) ProtoNet with BreakHis; (e) Meta DeepBDC with BreakHis; (f) CovNet with BreakHis.

To qualitatively assess our proposed approach's performance, we compare real and synthetic MRI images of prostate cancer in Fig. 4, representing each classification category. To the untrained eye, the synthetic images closely resemble the shape and features of the prostate, though they tend to show slightly more noise compared to real images. A comprehensive evaluation by a radiology expert is necessary to verify the presence of features associated with each classification. Nonetheless, our preliminary findings suggest that incorporating these synthetic images into the class prototype computation enhances classification performance, positioning our approach as a promising task-agnostic method for tumor characterization in data-limited scenarios. In future work, we will investigate how improvements in image quality and feedback from radiology experts can further boost the effectiveness of our approach.

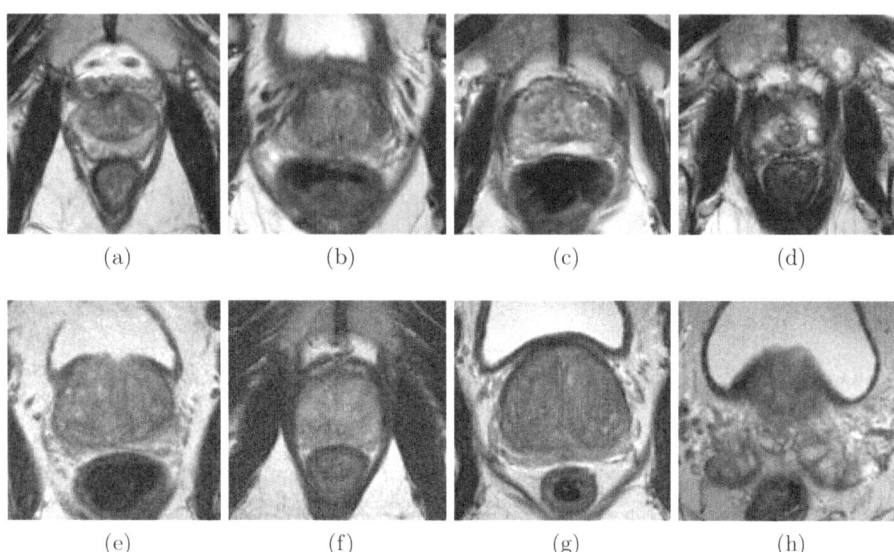

Fig. 4. Examples of real and synthetic MRI images of prostate cancer categorized by ISUP grade. (a) ISUP 2 real image; (b) ISUP 3 real image; (c) ISUP 4 real image; (d) ISUP 5 real image; (e) ISUP 2 synthetic image; (f) ISUP 3 synthetic image; (g) ISUP 4 synthetic image; (h) ISUP 5 synthetic image.

4 Conclusion

In this work, we combined the strengths of meta-learning and diffusion-based generative models to tackle tumor characterization in low-data scenarios. Our novel approach involves jointly training a diffusion model and a feature extractor within an episodic framework, utilizing the generation of synthetic support samples to enhance the creation of more informative prototypes. Additionally, we introduced a dynamic weighting mechanism that adjusts based on the similarity between generated samples and the original prototype of the same class, helping to mitigate the impact of generating poor or uninformative images, particularly during the early training stages. Preliminary experiments across various prototypical frameworks and two tumor characterization tasks demonstrate the effectiveness of our method in improving classification performance. Future work will focus on enhancing image generation quality and incorporating feedback from radiology experts to better assess the presence of class-related features in synthetic images.

Acknowledgements. This study has received funding from the European Union's Horizon 2020 research and innovation program under grant agreement No. 952159 (ProCAncer-I) and from the Regional Project PAR FAS Tuscany-NAVIGATOR. The funders had no role in the design of the study; the collection, analysis, and interpretation of data; or writing the manuscript.

Disclosure of Interests. The authors have no competing interests to declare that are relevant to the content of this article.

References

1. Cao, J., Yao, Z., Yu, L., Ling, B.W.K.: WPE: weighted prototype estimation for few-shot learning. Image Vis. Comput. **137**, 104757 (2023). https://doi.org/10.1016/j.imavis.2023.104757
2. Dai, Z., et al.: PFEMED: few-shot medical image classification using prior guided feature enhancement. Pattern Recogn. **134**, 109108 (2023)
3. Dosovitskiy, A., Brox, T.: Generating images with perceptual similarity metrics based on deep networks. Adv. Neural Inf. Process. Syst. **29** (2016)
4. Egevad, L., Delahunt, B., Srigley, J.R., Samaratunga, H.: International Society of Urological Pathology (ISUP) Grading of Prostate Cancer–An ISUP Consensus on Contemporary Grading (2016). https://doi.org/10.1111/apm.12533
5. Fei-Fei, L., Fergus, R., Perona, P.: One-shot learning of object categories. IEEE Trans. Pattern Anal. Mach. Intell. **28**(4), 594–611 (2006)
6. Fink, M.: Object classification from a single example utilizing class relevance metrics. Adv. Neural Inf. Process. Syst. **17** (2004)
7. Goodfellow, I., et al.: Generative adversarial nets. Adv. Neural Inf. Process. Syst. **27** (2014)
8. Guo, Z., Yan, Y., Yuan, Z., Yang, T.: Fast objective and duality gap convergence for non-convex strongly-concave min-max problems with PL condition. J. Mach. Learn. Res. **24**, 1–63 (2023)
9. He, F., Li, G., Si, L., Yan, L., Li, F., Sun, F.: Prototypeformer: learning to explore prototype relationships for few-shot image classification. arXiv preprint arXiv:2310.03517 (2023)
10. Ho, J., Jain, A., Abbeel, P.: Denoising diffusion probabilistic models. In: Larochelle, H., Ranzato, M., Hadsell, R., Balcan, M., Lin, H. (eds.) Advances in Neural Information Processing Systems, vol. 33, pp. 6840–6851. Curran Associates, Inc. (2020)
11. Ho, J., Salimans, T.: Classifier-free diffusion guidance. In: NeurIPS 2021 Workshop on Deep Generative Models and Downstream Applications (2021)
12. Jabbar, A., Li, X., Omar, B.: A survey on generative adversarial networks: variants, applications, and training. ACM Comput. Surv. **54**(8), 1–49 (2021)
13. Jiang, H., Gao, M., Li, H., Jin, R., Miao, H., Liu, J.: Multi-learner based deep meta-learning for few-shot medical image classification. IEEE J. Biomed. Health Inform. **27**(1), 17–28 (2022)
14. Kloeden, P.E., Platen, E.: Numerical Solution of Stochastic Differential Equations. Springer, Heidelberg (1992). https://doi.org/10.1007/978-3-662-12616-5
15. Ramesh, A., et al.: Zero-shot text-to-image generation. In: International Conference on Machine Learning, pp. 8821–8831. PMLR (2021)
16. Rezende, D.J., Mohamed, S., Wierstra, D.: Stochastic backpropagation and approximate inference in deep generative models. In: International Conference on Machine Learning, pp. 1278–1286. PMLR (2014)
17. Rombach, R., Blattmann, A., Lorenz, D., Esser, P., Ommer, B.: High-resolution Image Synthesis with Latent Diffusion Models (2021)
18. Saha, A., et al.: Artificial intelligence and radiologists at prostate cancer detection in MRI-the PI-CAI challenge. In: Medical Imaging with Deep Learning, Short Paper Track (2023)

19. Snell, J., Swersky, K., Zemel, R.: Prototypical networks for few-shot learning. In: Guyon, I., et al. (eds.) Advances in Neural Information Processing Systems, vol. 30. Curran Associates, Inc. (2017)
20. Sohl-Dickstein, J., Weiss, E., Maheswaranathan, N., Ganguli, S.: Deep unsupervised learning using nonequilibrium thermodynamics. In: Bach, F., Blei, D. (eds.) Proceedings of the 32nd International Conference on Machine Learning. Proceedings of Machine Learning Research, vol. 37, pp. 2256–2265. PMLR, Lille (2015)
21. Song, Y., Sohl-Dickstein, J., Kingma, D.P., Kumar, A., Ermon, S., Poole, B.: Score-based generative modeling through stochastic differential equations. In: International Conference on Learning Representations (2020)
22. Spanhol, F.A., Oliveira, L.S., Petitjean, C., Heutte, L.: A dataset for breast cancer histopathological image classification. IEEE Trans. Biomed. Eng. **63**(7), 1455–1462 (2015). https://doi.org/10.1109/TBME.2015.2496264
23. Wertheimer, D., Hariharan, B.: Few-shot learning with localization in realistic settings. In: Proceedings of the IEEE/CVF Conference on Computer Vision and Pattern Recognition, pp. 6558–6567 (2019)
24. Xie, J., Long, F., Lv, J., Wang, Q., Li, P.: Joint distribution matters: deep Brownian distance covariance for few-shot classification. In: Proceedings of the IEEE/CVF Conference on Computer Vision and Pattern Recognition, pp. 7972–7981 (2022)
25. Yuan, Z., Yan, Y., Sonka, M., Yang, T.: Large-scale robust deep AUC maximization: a new surrogate loss and empirical studies on medical image classification. In: Proceedings of the IEEE/CVF International Conference on Computer Vision, pp. 3040–3049 (2021). https://doi.org/10.1109/ICCV48922.2021.00303
26. Zhang, B., Li, X., Ye, Y., Huang, Z., Zhang, L.: Prototype completion with primitive knowledge for few-shot learning. In: Proceedings of the IEEE/CVF Conference on Computer Vision and Pattern Recognition, pp. 3754–3762 (2021)
27. Zhang, L., Zhou, F., Wei, W., Zhang, Y.: Meta-hallucinating prototype for few-shot learning promotion. Pattern Recogn. **136**, 109235 (2023). https://doi.org/10.1016/j.patcog.2022.109235
28. Zhu, Y., Cheng, Z., Wang, S., Zhang, H.: Learning de-biased prototypes for few-shot medical image segmentation. Pattern Recognit. Lett. **183**, 71–77 (2024)

Performance Evaluation of Deep Learning and Transformer Models Using Multimodal Data for Breast Cancer Classification

Sadam Hussain[1(✉)], Mansoor Ali[1], Usman Naseem[2],
Beatriz Alejandra Bosques Palomo[1], Mario Alexis Monsivais Molina[1],
Jorge Alberto Garza Abdala[1], Daly Betzabeth Avendano Avalos[4],
Servando Cardona-Huerta[4], T. Aaron Gulliver[3],
and Jose Gerardo Tamez Pena[4]

[1] Tecnologico de Monterrey, School of Sciences and Engineering, Mexico, Mexico
a01753094@tec.mx
[2] School of Computing, Macquarie University, Macquarie Park, Australia
[3] Department of Electrical and Computer Engineering, University of Victoria, Victoria, Canada
[4] School of Medical and Health Sciences, Tecnologico de Monterrey, Mexico, Mexico

Abstract. Rising breast cancer (BC) occurrence and mortality are major global concerns for women. Deep learning (DL) has demonstrated superior diagnostic performance in BC classification compared to human expert readers. However, the predominant use of unimodal (digital mammography) features may limit the current performance of diagnostic models. To address this, we collected a novel multimodal dataset comprising both imaging and textual data. This study proposes a multimodal DL architecture for BC classification, utilizing images (mammograms; four views) and textual data (radiological reports) from our new in-house dataset. Various augmentation techniques were applied to enhance the training data size for both imaging and textual data. We explored the performance of eleven SOTA DL architectures (VGG16, VGG19, ResNet34, ResNet50, MobileNet-v3, EffNet-b0, EffNet-b1, EffNet-b2, EffNet-b3, EffNet-b7, and Vision Transformer (ViT)) as imaging feature extractors. For textual feature extraction, we utilized either artificial neural networks (ANNs) or long short-term memory (LSTM) networks. The combined imaging and textual features were then inputted into an ANN classifier for BC classification, using the late fusion technique. We evaluated different feature extractor and classifier arrangements. The VGG19 and ANN combinations achieved the highest accuracy of 0.951. For precision, the VGG19 and ANN combination again surpassed other CNN and LSTM, ANN based architectures by achieving a score of 0.95. The best sensitivity score of 0.903 was achieved by the VGG16+LSTM. The highest F1 score of 0.931 was achieved by VGG19+LSTM. Only the VGG16+LSTM achieved the best area under the curve (AUC) of 0.937, with VGG16+LSTM closely following with a 0.929 AUC score.

Keywords: Breast Cancer · Feature Fusion · Multi-modal Classification · Deep Learning

1 Introduction

BC is the most prevalent disease among women [20]. Anticipated figures for 2023 suggest that the United States will likely witness 1,958,310 new cancer diagnoses and 609,820 cancer-related fatalities [8]. Digital mammography and other imaging modalities have long been used for BC detection [14]. However, due to the large volume of data, radiologists often struggle to process it in a timely manner. To assist them, various Computer-Aided Diagnosis (CAD) systems have been developed [3,25].

Traditionally, many CAD systems are tailored to process single-modality data, such as images, text, or audio. However, recent studies indicate that using multimodal data as input to SOTA DL models can yield promising results in the medical field, particularly in BC diagnosis and prognosis [7,13,19,22]. Multimodal data is crucial for BC diagnosis and prognosis, as it provides comprehensive information about various aspects such as radiomics characteristics, clinical features, and imaging features of the disease. Integrating different types of data allows DL methods to achieve better performance in BC classification, prognosis, and survival prediction than using only a single modality [13].

Multimodal BC classification has received increasing attention of researchers. Most approaches use imaging data along with textual reports for BC diagnosis [2,11] or some studies such as [11] used clinical factors. Some studies employed multidimensional data for predicting both short-term and long-term BC risk [19].

In this study, we provide a comparative analysis of various SOTA DL models with ViT [5]. We also collected a multimodal dataset comprising digital mammography (four views; L-CC, L-MLO, R-CC and R-MLO) and radiological reports. We propose a DL and transformer-based pipeline with different backbone networks to extract features from the multimodal data, applying a late fusion technique to combine imaging and textual features before training a classifier for BC classification. We also evaluated the performance of the transformer-based architecture, ViT, which has shown promising outcomes on text and imaging data [6].

The proposed model uses CNNs and ViT for feature extraction from digital mammography and either ANN or LSTM for feature extraction from radiological reports. The fused features are then assigned to an ANN classifier for the final classification of benign or malignant cancer. We observed that models combining LSTM for textual feature extraction generally performed better than those using simple ANN. Furthermore, it has been observed that VGG16 and VGG19 models outperformed other SOTA DL architectures and ViT across all the evaluated metrics The proposed multimodal pipeline is illustrated in Fig. 1.

2 Proposed Work

In this work, we evaluate the performance of various DL and transformer based architectures on a newly collected multimodal dataset for BC diagnosis.

The key contributions of this work are highlighted as follows:

i. A new in-house dataset comprising multi-view mammograms and radiological reports has been collected and preprocessed.

ii. A DL and transformer-based multimodal pipeline for BC classification is employed on a new in-house dataset. In this pipeline, SOTA DL models, such as VGG16 [18], VGG19 [18], ResNet34 [9], ResNet50 [9], MobileNet-v3 [12], EffNet-b0 [21], EffNet-b1 [21], EffNet-b2 [21], EffNet-b3 [21], EffNet-b7 [21], and a transformer-based model, ViT [4], are used to extract imaging features, with either LSTM or ANN employed to extract textual features.

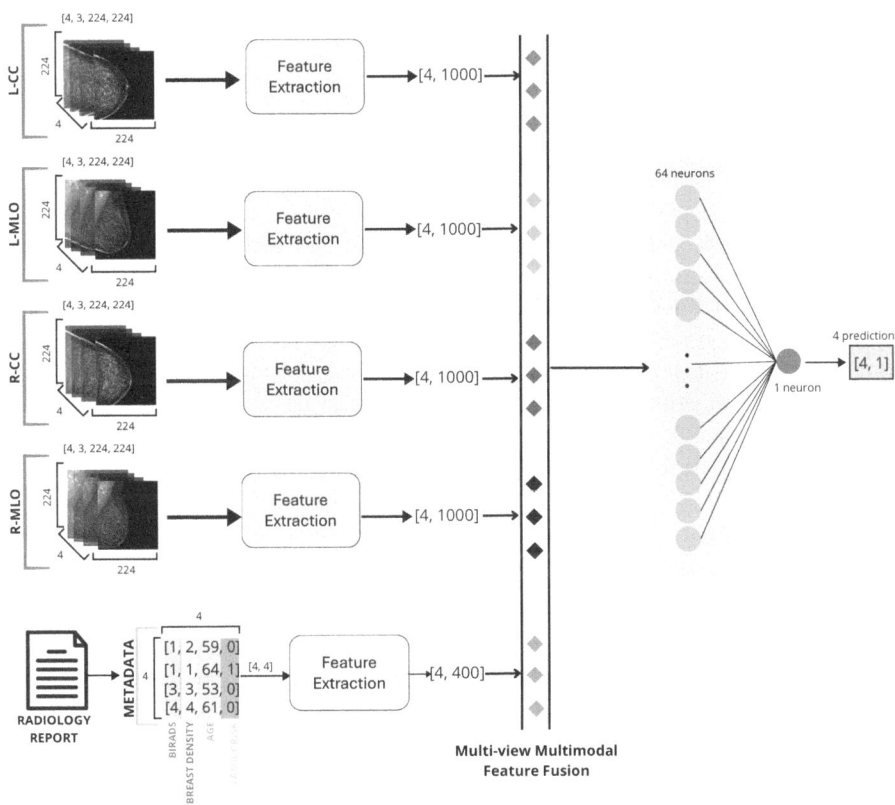

Fig. 1. Proposed Multi-Modal Model for Processing Multi-Dimensional Data

3 Methodology

This section highlights the methodology employed in this study, encompassing data collection, preprocessing, data augmentation, as well as the training and testing of data. It also details the multimodal architecture utilized, implementation specifics, classification, and evaluation of the models.

3.1 Data

The dataset used in this investigation was provided by the breast radiology department at TecSalud Hospital in Monterrey, Mexico. In addition to digital mammograms, it includes digital breast tomosynthesis (DBT), digital mammography, and ultrasound (US) based radiological reports. Anonymized reports and mammograms were used to ensure anonymity. The reports, which were gathered between January and December 2014 and were initially written in Spanish, were translated into English using Google Translate. Bilingual radiologists then confirmed the translation. Two seasoned radiologists wrote and evaluated each report. One radiologist reviewed the translated reports. In cases where the first radiologist disagreed with the translation, the second radiologist revised and finalized the translation. There were a thousand examples in the dataset at first. Four views (LCC, LMLO, RCC, and RMLO) and related textual data (BIRADS categories, breast density scores, patient ages, family cancer histories, and lesion laterality) make up each patient's data set. Lesion laterality reveals if a patient has had cancer in one or both breasts. Three values are used to characterize it: 0 for left breast, 1 for right breast, and 2 for cancer in both breasts. Laterality was only performed to determine whether the patient had malignancy. We used 770 cases with a total of 3080 mammograms and related metadata after preprocessing, which comprised deleting reports with missing values, duplicates, and BIRADS categories 0 (inconclusive) and 6 (biopsy-proven). There are exactly equal numbers of positive (385) and negative (385) cases in the dataset. The final model building and evaluation involved women in all cases, with an average patient age of 53.

3.2 Data Preprocessing

The original dataset, in Spanish, consisted of 7,904 samples. After translation, we removed redundant data, missing values, and cases with BIRADS scores of 0 or 6. Following preprocessing, we collected 5,046 samples. Each patient entry included a brief clinical indication of the study, personal risk history, imaging description, findings, and diagnostic impression. From these radiological reports, we extracted BIRADS categories (1–5), breast density scores, patients' ages, and family history of cancer, which were ultimately used for model building.

3.3 Data Augmentation

The augmentation techniques applied to the mammogram dataset were carefully selected to introduce diversity and robustness into the training data. Horizontal and vertical flipping were utilized to mirror the images, exposing the model to different orientations. Furthermore, the implementation of random resized crops, targeting an output size of 224 × 224 pixels, aimed to instill scale and translation invariance during model training. Dynamic transformations, such as shifts, scales, and rotations, were also incorporated, allowing for rotations up to 90°C and scale adjustments within the range of 0.8 to 1.2. This diverse set of augmentations serves the crucial purpose of expanding the effective size of the training dataset. By exposing the model to varied perspectives, the augmentation strategy becomes instrumental in preventing overfitting and enhancing the model's adaptability to diverse input scenarios, ultimately contributing to better performance.

3.4 Multimodal Deep Architecture

We introduced a multimodal architecture to classify BC instances as positive or negative. The proposed architecture consists of independent DL and transformer-based architectures (pre-trained on ImageNet "IMAGENET1K_V1" weights provided by the torchvision library) as a backbone for extracting imaging features from mammograms. The DL architectures used in this study are variants of VGG, ResNet, EffNet and MobileNetv3. We chose these techniques because of their SOTA performance on natural and medical image classification tasks [4,15,17,21,24]. For extracting textual features that were extracted from radiological reports as tabular data we used simple ANN or long short-term memory (LSTM). These simple architectures perform better on tabular data. Afterwards, the feature vectors extracted from both modalities were fused before the final classification task. This vector is then passed through a linear classification layer for the prediction of malignant or benign cancer.

Let $\mathcal{I}$ represents mammogram input, where there are four input images one for each view of the mammogram, therefore, i_1, i_2, i_3 and i_4, represents each view of the mammogram respectively. Features extracted from the four mammogram views are represented as f_1, f_2, f_3 and f_4. Let $\mathcal{T}$ represents input tabular data extracted from radiological reports. Features extracted from tabular data are shown as f_t. Therefore, the concatenated feature vector represented as $\mathcal{F}_c = $ concat(f_1, f_2, f_3, f_4, f_t). The $\mathcal{F}_c$ is then fed to a linear layer for BC classification.

The transformer architecture is an attention-based encoder-decoder model. The encoder processes the input, and the decoder generates the output. Importantly, the transformer architecture does not depend on recurrence or convolutions to produce an output. The architecture employs stacked self-attention and point-wise, fully connected layers for both the encoder and decoder, as detailed in Vaswani et al., 2017 [23].

The self-attention mechanism is one of the most important parts of the transformer architecture. It makes the model to consider the relationships between all positions in the input simultaneously with:

$$MultiHead(Q, K, V) = Concat(head_1, ..., head_h)W^O \quad (1)$$

$$head_i = Attention(QW_i^Q, KW_i^K, VW_i^V), \quad (2)$$

where Q is the query, K denotes to the key, and V is the value [23].

The ViT is an extension of the transformer architecture, utilizing only the encoder part. The input image is divided into fixed-size, non-overlapping patches. Each patch is linearly embedded to form vectors, with an added "classification token." Thus, the transformer's input is the classification token combined with the patch embeddings. The transformer's output then serves as the input for a classification head.

3.5 Implementation Details

For training and testing all models, we utilized an NVIDIA RTX A4000 GPU. This GPU is equipped with 16 GB of memory and supports PCI Express Gen 4. Additionally, it features third-generation tensor CUDA cores. Our network architecture incorporated ResNet34, ResNet50, VGG16, VGG19, MobileNet_v3, ViT, and EffNet_b0 as backbones for image encoding, and either ANN or LSTM for text encoding. The model weights were initialized from scratch rather than using pretrained models. We employed Adam as the optimizer to minimize loss, with a learning rate set at 0.0005. The batch size was fixed at 8, and the models were trained over 100 epochs. The Rectified Linear Unit (ReLU) was used as the activation function, with a dropout rate of 0.2. The dataset was split into training and validation sets at a ratio of 80% to 20%, respectively. This data split was consistent across all experiments. Model performance was evaluated using the validation set.

4 Results

In our comprehensive evaluation, we used a variety of performance metrics to assess the effectiveness of our multimodal models, ensuring a thorough understanding of their classification capabilities. These metrics included Accuracy, Precision, Sensitivity, F1 Score, and AUC. Our results demonstrated significant achievements across different metrics, highlighting the distinct strengths of various models.

The highest accuracy, at 0.951, was achieved by the VGG19+ANN architecture, indicating the superior ability to correctly classify instances within the dataset. This was closely followed by the joint models of VGG19+LSTM, scoring 0.919. For precision, the VGG19 paired with ANN achieved the highest score of 0.95, surpassing other combined model configurations, followed by the ViT and LSTM combination scoring 0.932. This highlights VGG19's proficiency in minimizing false positives and ensuring precise classifications. In terms of sensitivity, the VGG16 and LSTM combination recorded the highest rate at 0.903, followed by VGG19+LSTM, at 0.900. This indicates their effectiveness in correctly identifying a significant proportion of true positive cases, crucial in applications where sensitivity is key. The highest F1 Score, reflecting a balance between precision and recall, was 0.931 using the VGG19+LSTM, with the VGG19+ANN close behind at 0.922. The AUC analysis showed the VGG16+LSTM model leading with a score of 0.937, demonstrating its ability to differentiate between positive and negative cases effectively, followed by the VGG16 and ANN combination at 0.929. The results of combining SOTA DL architectures with LSTM for BC classification are presented in Table 1, and those with ANN in Table 2.

The least performing model was the ResNet34 combined with ANN, recording an accuracy of 0.692. The MobileNet_v3 combined with ANN was the least accurate in the LSTM category, at 0.813. The EffNet_b3 paired with ANN had the lowest precision of 0.711, while ResNet50 with LSTM scored the lowest at 0.747. In terms of sensitivity, the lowest score, 0.643, was recorded by the MobileNet_v3 and ANN combination, followed by the EffNet_b3 combined with ANN scoring 0.653. The EffNet_b3 and ANN combination had the lowest F1 Score at 0.652, while the lowest in the LSTM category was 0.723 for EffNet_b0, EffNet_b1, EffNet_b2. Lastly, the lowest AUC score of 0.678 was achieved by the MobileNet_v3 and ANN combination, and 0.679 by the ViT and LSTM combination.

Table 1. Overview of the Results of Different DL Models for BC Classification

Feature Extractor	Classifier	ACC	Precision	Sensitivity	F1 Score	AUC
VGG16+LSTM	ANN	0.882	0.902	**0.903**	0.902	**0.937**
VGG19+LSTM	ANN	**0.919**	0.923	0.900	**0.931**	0.907
ResNet34+LSTM	ANN	0.910	0.782	0.899	0.753	0.739
ResNet50+LSTM	ANN	0.889	0.747	0.899	0.753	0.798
MobileNet_v3+LSTM	ANN	0.813	0.835	0.753	0.903	0.679
EffNet_b0+LSTM	ANN	0.834	0.769	0.892	0.723	0.757
EffNet_b1+LSTM	ANN	0.834	0.772	0.895	0.723	0.759
EffNet_b2+LSTM	ANN	0.842	0.893	0.876	0.723	0.807
EffNet_b3+LSTM	ANN	0.827	0.758	0.903	0.736	0.719
EffNet_b7+LSTM	ANN	0.825	0.884	0.872	0.738	0.685
ViT+LSTM	ANN	0.893	**0.932**	0.863	0.871	0.925

5 Discussion

The data in healthcare is inherently multimodal [1]. It includes scans in the form of images, doctors' analyses as notes or radiological reports, and audio data, such as ECG/EEG, which provide insights for diagnosis, prognosis, and treatment decisions [16]. However, most healthcare studies focus on a single modality of data (image, text, or audio) due to the scarcity of multimodal data, often yielding less than optimal results. Since the last decade, researchers have been combining different data modalities to achieve better outcomes. Current methods, though,

Table 2. Overview of the Results of Different DL Models for BC Classification

Feature Extractor	Classifier	ACC	Precision	Sensitivity	F1 Score	AUC
VGG16+ANN	ANN	0.900	0.912	**0.893**	0.897	**0.929**
VGG19+ANN	ANN	**0.951**	**0.950**	0.884	**0.922**	0.915
ResNet34+ANN	ANN	0.692	0.731	0.716	0.692	0.748
ResNet50+ANN	ANN	0.721	0.752	0.705	0.681	0.723
MobileNet_v3+ANN	ANN	0.754	0.858	0.643	0.668	0.678
EffNet_b0+ANN	ANN	0.781	0.714	0.658	0.668	0.709
EffNet_b1+ANN	ANN	0.781	0.714	0.658	0.668	0.703
EffNet_b2+ANN	ANN	0.782	0.887	0.825	0.851	0.828
EffNet_b3+ANN	ANN	0.781	0.711	0.653	0.652	0.719
EffNet_b7+ANN	ANN	0.781	0.812	0.658	0.687	0.689
ViT+ANN	ANN	0.871	0.903	0.819	0.848	0.844

still fall short in terms of generalizability and accuracy comparable to that of physicians [10].

In this work, we extracted features from images and textual data, combining them for the final classification of benign and malignant cancer. We employed various SOTA DL and transformer-based architectures for feature extraction from mammograms and radiological reports. These included VGG16, VGG19, ResNet34, ResNet50, MobileNet_v3, EffNet_b0, EffNet_b1, EffNet_b2, EffNet_b3, EffNet_b7, and ViT for mammogram feature extraction. For textual feature extraction, we used ANN or LSTM techniques. We then merged these features using a late fusion strategy and employed an ANN architecture for classification. ViT, in combination with LSTM or ANN, couldn't achieve better or comparable results to other SOTA DL models in terms of accuracy, precision, sensitivity, F1 score and AUC. However, ViT+LSTM ranked second highest in the AUC curve with a score of 0.925, just behind VGG16+LSTM with the highest score of 0.937.

We observed that using LSTM for textual feature extraction with any DL architecture as a backbone for imaging feature extraction can slightly enhance the overall model's performance in terms of sensitivity, F1 score and AUC, compared to basic ANN. However, ANN combinations outperformed LSTM in metrics like accuracy and precision. Both ANN and LSTM effectively utilized the rich contextual information in textual data, improving the model's capability to discern complex patterns and relationships. It is also observed that SOTA DL architectures performed better than transformer based architecture ViT. In addition, across all the evaluated metrics, VGG based architectures performed better than any other architecture used in this study.

6 Conclusion

The aim of this study was to collect and preprocess a new dataset of digital mammograms and radiological reports and give a baseline of SOTA DL and transformer-based architecture combined with either ANN or LSTM using a multimodal fusion approach. The highest accuracy was achieved by the VGG19 model in conjunction with ANN. For precision, the top score was again attained by VGG19 associated with ANN, followed by the combined model of ViT and LSTM. The best sensitivity score was recorded by the VGG16 with LSTM, followed by the combined models of VGG19 and LSTM fused model. The highest F1 scores were achieved by VGG19 in conjunction with LSTM, followed by the joint model of VGG19 and ANN. The best AUC score was achieved by VGG16 and LSTM combination. The VGG16 and ANN combination secured the second best AUC score. Our observations indicated that models combining LSTM with DL architectures either performed better or comparably to those using ANN in various metrics. The overall performance of the joint ViT model with LSTM or ANN was less as compared to SOTA DL architectures in most of the evaluation metrics. The results suggest that among the five evaluation metrics, SOTA DL models performed better in all evaluation metrics, while the transformer architecture along with LSTM was able to achieve second best in a precision metric only. In addition, the VGG16 and VGG19 outperformed all the SOTA DL architectures and ViT across all metrics which suggest the ability of VGG architectures in classifying the positive and negative BC instances across all the evaluation metrics. These findings further indicate that incorporating metadata extracted from radiological reports alongside images can enhance the model's performance in predicting BC. This approach has potential applications in various medical fields due to the inherently multimodal nature of healthcare data.

Acknowledgements. The authors would like to thank the Tecnológico de Monterrey and CONAHCYT for supporting their studies.

References

1. Acosta, J.N., Falcone, G.J., Rajpurkar, P., Topol, E.J.: Multimodal biomedical AI. Nat. Med. **28**(9), 1773–1784 (2022)
2. Akselrod-Ballin, A., et al.: Predicting breast cancer by applying deep learning to linked health records and mammograms. Radiology **292**(2), 331–342 (2019)
3. Calisto, F.M., Santiago, C., Nunes, N., Nascimento, J.C.: Breastscreening-AI: evaluating medical intelligent agents for human-AI interactions. Artif. Intell. Med. **127**, 102285 (2022)
4. Dosovitskiy, A., et al..: An image is worth 16x16 words: transformers for image recognition at scale. In: International Conference on Learning Representations (ICLR) (2021). https://openreview.net/forum?id=YicbFdNTTy
5. Dosovitskiy, A., et al.: An image is worth 16x16 words: transformers for image recognition at scale. arXiv preprint arXiv:2010.11929 (2020)

6. Fields, C., Kennington, C.: Vision language transformers: a survey. arXiv e-prints pp. arXiv–2307 (2023)
7. Gao, J., et al.: Integrative analysis of complex cancer genomics and clinical profiles using the cbioportal. Sci. Signal. **6**(269), pl1 (2013)
8. Hansebout, R.R., Cornacchi, S.D., Haines, T.A., Goldsmith, C.H.: How to use an article about prognosis. Canadian J. Surg. J. Canadien Chirurgie **52 4**, 328–336 (2009)
9. He, K., Zhang, X., Ren, S., Sun, J.: Deep residual learning for image recognition. In: Proceedings of the IEEE Conference on Computer Vision and Pattern Recognition, pp. 770–778 (2016)
10. Heiliger, L., Sekuboyina, A., Menze, B., Egger, J., Kleesiek, J.: Beyond medical imaging-a review of multimodal deep learning in radiology. TechRxiv (19103432) (2022)
11. Holste, G., Partridge, S.C., Rahbar, H., Biswas, D., Lee, C.I., Alessio, A.M.: End-to-end learning of fused image and non-image features for improved breast cancer classification from MRI. In: Proceedings of the IEEE/CVF International Conference on Computer Vision, pp. 3294–3303 (2021)
12. Howard, A., et al.: Searching for mobilenetv3. In: Proceedings of the IEEE/CVF International Conference on Computer Vision, pp. 1314–1324 (2019)
13. Huang, S.C., Pareek, A., Zamanian, R., Banerjee, I., Lungren, M.P.: Multimodal fusion with deep neural networks for leveraging ct imaging and electronic health record: a case-study in pulmonary embolism detection. Sci. Rep. **10**(1), 22147 (2020)
14. Hussain, S., Lafarga-Osuna, Y., Ali, M., Naseem, U., Ahmed, M., Tamez-Peña, J.G.: Deep learning, radiomics and radiogenomics applications in the digital breast tomosynthesis: a systematic review. BMC Bioinformatics **24**(1), 401 (2023)
15. Ikechukwu, A.V., Murali, S., Deepu, R., Shivamurthy, R.: Resnet-50 vs vgg-19 vs training from scratch: a comparative analysis of the segmentation and classification of pneumonia from chest x-ray images. Global Transit. Proc. **2**(2), 375–381 (2021)
16. Pei, X., Zuo, K., Li, Y., Pang, Z.: A review of the application of multi-modal deep learning in medicine: bibliometrics and future directions. Int. J. Comput. Intell. Syst. **16**(1), 44 (2023)
17. Qian, S., Ning, C., Hu, Y.: Mobilenetv3 for image classification. In: 2021 IEEE 2nd International Conference on Big Data, Artificial Intelligence and Internet of Things Engineering (ICBAIE), pp. 490–497. IEEE (2021)
18. Simonyan, K., Zisserman, A.: Very deep convolutional networks for large-scale image recognition. arXiv preprint arXiv:1409.1556 (2014)
19. Sun, D., Wang, M., Li, A.: A multimodal deep neural network for human breast cancer prognosis prediction by integrating multi-dimensional data. IEEE/ACM Trans. Comput. Biol. Bioinf. **16**(3), 841–850 (2018)
20. Sung, H., et al.: Global cancer statistics 2020: Globocan estimates of incidence and mortality worldwide for 36 cancers in 185 countries. CA: Cancer J. Clin. **71**(3), 209–249 (2021)
21. Tan, M., Le, Q.: Efficientnet: rethinking model scaling for convolutional neural networks. In: International Conference on Machine Learning, pp. 6105–6114. PMLR (2019)
22. Tomczak, K., Czerwińska, P., Wiznerowicz, M.: Review the cancer genome atlas (TCGA): an immeasurable source of knowledge. Contemp. Oncol./Współczesna Onkologia **2015**(1), 68–77 (2015)
23. Vaswani, A., et al.: Attention is all you need. Adv. Neural Inf. Process. Syst. **30** (2017)

24. Xu, W., Fu, Y.L., Zhu, D.: Resnet and its application to medical image processing: research progress and challenges. Comput. Methods Prog. Biomed. **240**, 107660 (2023)
25. Yassin, N.I., Omran, S., El Houby, E.M., Allam, H.: Machine learning techniques for breast cancer computer aided diagnosis using different image modalities: a systematic review. Comput. Methods Prog. Biomed. **156**, 25–45 (2018)

Detection and Segmentation

On Undesired Emergent Behaviors in Compound Prostate Cancer Detection Systems

Erlend Sortland Rolfsnes[1], Philip Thangngat[1], Trygve Eftestøl[1], Tobias Nordström[2,3], Fredrik Jäderling[4,5], Martin Eklund[2], and Alvaro Fernandez-Quilez[1,2(✉)]

[1] Department of Electrical Engineering and Computer Science, University of Stavanger, Stavanger, Norway
[2] Department of Medical Epidemiology and Biostatistics, Karolinska Institutet, Solna, Sweden
alvaro.f.quilez@uis.no
[3] Department of Clinical Sciences, Danderyd Hospital, Danderyd, Sweden
[4] Department of Radiology, Capio Saint Göran Hospital, Stockholm, Sweden
[5] Department of Molecular Medicine and Surgery, Karolinska Institutet, Solna, Sweden

Abstract. Artificial intelligence systems show promise to aid in the diagnostic pathway of prostate cancer (PC), by supporting radiologists in interpreting magnetic resonance images (MRI) of the prostate. Most MRI-based systems are designed to detect clinically significant PC lesions, with the main objective of preventing over-diagnosis. Typically, these systems involve an automatic prostate segmentation component and a clinically significant PC lesion detection component. In spite of the compound nature of the systems, evaluations are presented assuming a standalone clinically significant PC detection component. That is, they are evaluated in an idealized scenario and under the assumption that a highly accurate prostate segmentation is available at test time. In this work, we aim to evaluate a clinically significant PC lesion detection system accounting for its compound nature. For that purpose, we simulate a realistic deployment scenario and evaluate the effect of two non-ideal and previously validated prostate segmentation modules on the PC detection ability of the compound system. Following, we compare them with an idealized setting, where prostate segmentations are assumed to have no faults. We observe significant differences in the detection ability of the compound system in a realistic scenario and in the presence of the highest-performing prostate segmentation module (DSC: 90.07 ± 0.74), when compared to the idealized one (AUC: 77.97 ± 3.06 and 84.30 ± 4.07, $P<.001$). Our results depict the relevance of holistic evaluations for PC detection compound systems, where interactions between system components can lead to decreased performance and degradation at deployment time.

Keywords: MRI · Compound Systems · Prostate Cancer · Deep Learning

1 Introduction

Over the last decade, magnetic resonance imaging (MRI) has become an important tool for prostate cancer (PC) detection, staging and treatment planning [4]. Its surge in relevance for PC diagnosis is expected to substantially increase the amount of imaging examinations, and, consequently, the radiologist workload [7,11]. The increasing volume of MRI coupled with a shortage of specialists, presents a favorable scenario to delegate time-consuming tasks such as the detection of clinically significant PC (csPC) lesions to artificial intelligence (AI) systems [18].

Whilst AI systems for csPC detection hold potential to augment the radiologist abilities, reduce reader variability and shorten the study times, their availability is not translated into clinical adoption [1,24]. Limitations in adopting these systems are often reported to stem from a misalignment between system evaluation and its intended use in clinical practice [12]. This misalignment is typically exemplified by AI systems exhibiting an overoptimistic performance during their development that does not match their performance at deployment time [10,23,24].

AI systems for csPC detection are typically compound systems that integrate different interacting components [18]. In particular, most csPC detection systems rely on an AI prostate gland segmentation component followed by a csPC detection module that depends on the quality of the previously segmented prostate [22,23]. In spite of the dependence and expected interaction of the components, the evaluation of the compound system is performed assuming a standalone detection module. That is, the compound system is evaluated in an idealized scenario and under the assumption of the availability of a highly accurate prostate segmentation [22]. Hence, without accounting for the emergent behaviors due to the components' interaction and resulting in an evaluation that is not representative of the system as a whole.

Against this background, we aimed at performing a holistic evaluation of a typical compound csPC detection system that integrates a prostate segmentation module and a detection module [8,18]. Further, we aim to investigate emergent behaviors that stem from unexpected results that propagate from the prostate segmentation module to the csPC detection module (Fig. 1). We hypothesized that unaccounted emergent behaviors stemming from the interaction between modules might be associated with an overoptimistic report in the performance of the system.

2 Related Work

Multiple works have attempted to develop csPC detection compound systems. In the work by Cao *et al.* [2], the authors propose a compound system based on a 2D network for csPC detection and lesion grade prediction, reaching a 0.81 area under the curve (AUC) detection ability for MRI slices that contained lesions. In the work of Sanyal et al. [21], the authors proposed a compound system where the first component segments the prostate gland, whilst the second component performs csPC detection based on the previously segmented prostate (AUC = 0.86). More recently, Sanford et al. [20], developed a system that relies on previously segmented prostates to detect csPC from MRI. In the work, they compare the performance of the system with expert radiologists reaching a 56% agreement. Fernandez-Quilez et al. [8] proposed a compound system that detects PC by obtaining a surrogate biomarker of age from MRI (AUC = 0.981). In this work, the csPC detection module assumes the availability of high quality prostate segmentation masks. Finally, Saha et al. [18] present a compound csPC detection system based on a prostate segmentation module and an ensemble of different architectures based on the largest data sample to date for csPC detection (AUC = 0.91). The system outperformed a consensus of 62 radiologist specialists. Although these methods achieve a high performance, they all share a lack of evaluation in terms of interaction between the modules that compose the proposed compound systems.

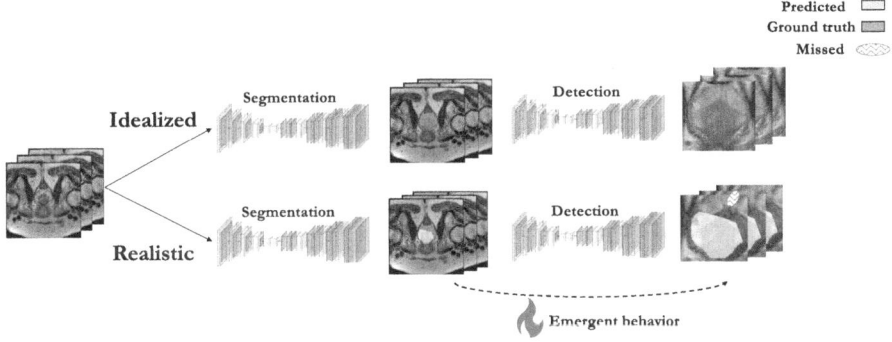

Fig. 1. Compound clinically-significant prostate cancer detection system evaluated under the assumption that no faults will be present in the output of the segmentation module (idealized), and accounting for the behaviors emerging from the interaction of the segmentation and detection modules (realistic).

3 Materials and Methods

3.1 Study Cohort

In our retrospective single-center study, we leverage data from the open access ProstateX challenge data (Radboud University, Netherlands) [15]. ProstateX consists of a collection of prostate MRI exams collected with the purpose of validating modern AI algorithms for the diagnostic classification of csPC. Subjects were recruited on the basis of suspicion of PC based on high PSA levels. All subjects had PC diagnosis confirmed through an MRI-guided biopsy [14].

The ProstateX challenge cohort consisted of 204 subjects (median age 66 years [range, 48–83]), with available T2-weighted (T2w) spin sequences and diffusion-weighted (DW) magnetic resonance exams. The cohort had pixel-level annotations for the prostate gland and prostate lesions obtained by two experienced board-certified radiologists with > 5 years of experience [3]. All annotations were obtained with ITK-SNAP v.380 software (http://www.itksnap.org/) [25].

Magnetic Resonance Imaging. From the original 204 available subjects, we leverage 200 subjects 3.0 T (T) axial T2w spin sequences based on the availability of both pixel-level lesion annotations and prostate gland annotations (Fig. 2). Images were acquired with either Siemens (Siemens Health Engineers, Erlangen, Germany) or Philips (Philips & Co, Eindhoven, The Netherlands) scanners. The T2w MRI exams had an in-plane resolution of [0.5–0.562 mm] × [0.5–0.562 mm] × [3.0–3.15 mm].

Preprocessing and Splitting. To have a common space of reference, we apply a re-sampling by linear interpolation to all the T2w sequences to a resolution of $0.5 \times 0.5 \times 3.6$ mm. In addition, we also apply N4 bias field correction filtering and normalization of the T2w exams pixel intensity [13]. Following, we split the cohort in 80/20% whilst avoiding cross-contamination. The splitting resulted in 160 and 40 patients for the train and test set, respectively. The splitting is performed stratifying by PC lesion type, which results in 498 (72.80%) non-clinically significant PC (non-csPC) lesions in the train set, 186 (27.20%) in the test set, 222 (71.61%) csPC lesions in the train set and 88 (28.39%) in the test set.

3.2 Compound PC Detection System

As shown in Fig. 2, the compound system integrates two modules: automatic prostate segmentation [9,13] and csPC lesion detection [8,18]. We train and evaluate both modules with the same data split, to avoid undesired data leakage.

Prostate Segmentation Module. We leverage two different segmentation networks. In the first case, we focus on a multi-view segmentation network with a good generalization ability based on previously reported results [13]. In addition, we employ a 2D nnU-Net based on its wide adoption and positive results in previous prostate segmentation challenges [9]. The training follows the steps described in the original article for the multi-view network and the default configuration for the 2D nnU-Net, respectively [9,13]. Given the objectives of this work, exploration of architectural or training modifications is considered out of the scope of the work. Training and evaluation is performed on an NVIDIA A100 GPU (NVIDIA Corporation, Santa Clara, USA).

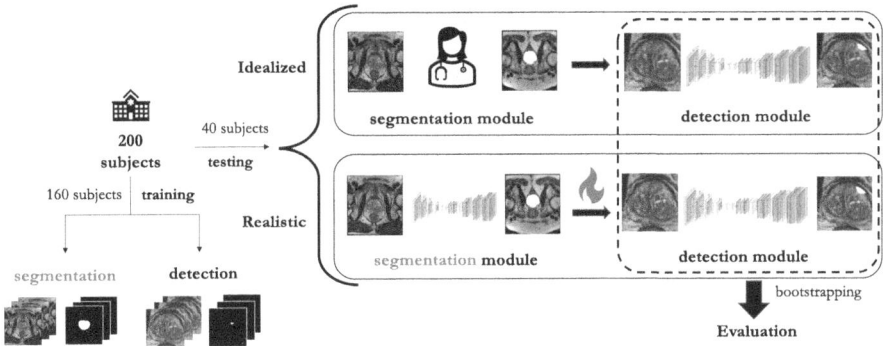

Fig. 2. Technical development of the project, depicting the splitting of the data, MRI exams and ground truth used to train the segmentation and detection module and their test protocol. Whilst in the idealized case it is assumed that the segmentation is absent of faults, in the realistic case we explore emergent behaviors stemming from segmentation faults.

Clinically Significant PC Detection Module. Based on previously reported results [18], we adopt a standard U-Net architecture [17]. We use an ImageNet pre-trained model and a channel-wise convolutional block to map the original gray-scale images (1 channel) to the expected Red-Green-Blue (RGB) format (3 channels). Following standard approaches, we train the architecture using cropped 2D T2w slices as an input. We crop the slices by leveraging the prostate segmentation mask obtained from the previous module. The cropping is performed around the prostate gland with a 20 pixel margin to avoid edge cases and ensure that we capture the whole region of interest. We train using a 5-fold cross-validation (CV), resulting in 5 models. These models were then ensembled with equal weighting into a single system. Training and evaluation is performed on an NVIDIA A100 GPU (NVIDIA Corporation, Santa Clara, USA).

3.3 Emergent Behaviors

Following standard practices for compound PC systems, we start by providing a standalone evaluation of the csPC detection module at test time. That is, the cropped T2w input of the csPC detection module is obtained by leveraging the prostate segmentation ground truth. This is referred as idealized scenario, where the interaction between modules is disregarded. Following, we evaluate the modules accounting for possible emergent behaviors due to their interaction. This is referred as realistic scenario. In this case, we start by characterizing the performance of the segmentation module. Next, we characterize the performance of the csPC detection module by leveraging the output of the segmentation module as input for the csPC detection module. The realistic evaluation process is repeated for the multi-view and nnU-Net architectures, with heterogeneous segmentation performances (Fig. 1).

In both scenarios and for both modules, the results are presented at the patient level. That is, the results are presented for the whole patient exam (3D). For the segmentation module, the results are presented in terms of Dice Score Coefficient (DSC, %), Relative Volume Difference (RVD, %) and Hausdorff Distance (HD, mm). For the csPC detection module, the results are presented in terms of DSC and AUC. For the AUC, lesions that shared a minimum DSC of 0.10 when compared with the ground truth were considered true positives. We follow similar criteria to the one presented in previous works [16,19].

3.4 Statistical Testing

We obtain an estimate of the test results metrics for the compound system using a non-parametric bootstrapping with $n = 1000$ replicates without repetition. We assess statistical significance for AUC through permutation tests. A $P < 0.05$ is considered statistically significant. All results are presented as mean ± standard deviation (SD). All analyses were performed in Python 3 (www.python.org/downloads) with the open-sourced statsmodels 0.14.0 module (www.statsmodels.org).

Table 1. Results for the segmentation module when independently evaluated. Results are presented for two different segmentation architectures.

Architecture	DSC (%) ↑	RVD (%) ↓	HD (mm) ↓
nnU-Net [9]	85.98 ± 6.18	3.73 ± 9.02	7.76 ± 4.82
tU-Net [13]	90.07 ± 0.74	2.01 ± 1.30	1.76 ± 0.39

↑ larger better ↓ lower better, DSC = Dice Score Coefficient, RVD = Relative Volume Difference, HD = Hausdorff distance.

4 Results and Discussion

Table 1 depicts the experimental results for the segmentation module. We can observe that nnU-Net presents a DSC that is close to expert level radiologist annotation accuracy (DSC >= 0.86) [5], whilst tU-Net presents a DSC that is above that level. In terms of RVD and HD, we can observe that tU-Net presents a higher accuracy in volumetric estimation and a lower HD.

Table 2 presents the results for the compound system in an idealized and realistic scenarios, respectively. When analysing the results of the detection module as a standalone module in the idealized scenario, the compound system presents a bootstrapped AUC = 84.30 ± 4.07 for csPC lesions. When evaluated accounting for emergent behaviors stemming from the interaction between the segmentation module and the detection module, we can observe a significant decrease in the AUC performance. Specifically, when leveraging nnU-Net AUC = 71.30 ±

3.83 and when leveraging tU-Net AUC = 77.97 ± 3.06. The results show that a higher individual performance in the first module (tU-Net) is translated into a higher compound system performance.

As depicted qualitatively in Fig. 3, we can observe from an intuitive perspective the effect of a faulty segmentation for the input of the detection module. In particular, we can observe the significant effect in the input of the detection module of the different segmentation module output architectures.

Our retrospective study provides some degree of evidence of the emergence of undesired behaviors when accounting for the interaction of modules in a typical compound PC detection system. In particular, for a compound system that leverages a segmentation module and a csPC detection module based on U-Net variants [13,18,21]. Our results support the hypothesis that unaccounted interactions are correlated with an overoptimistic report in the compound system performance.

Table 2. Results for the compound system when evaluated in an idealized setting and in a realistic setting. In the first case, the detection module is evaluated as a standalone module, without accounting for faulty segmentations. In the second case, emergent behaviors are taken into account by using the output of the segmentation module as the input for the detection one.

Scenario	Architecture	Lesion	DSC (%) ↑	AUC (%) ↑	P_{AUC}
Idealized	×	csPC	12.40 ± 1.96	84.30 ± 4.07	×
		non-csPC	16.97 ± 1.34	86.90 ± 2.23	×
Realistic	nnU-Net [9]	csPC	7.14 ± 2.05	71.30 ± 3.83	<.001[†]
		non-csPC	9.84 ± 0.98	75.57 ± 3.61	<.001[†]
	tU-Net [13]	csPC	9.24 ± 3.01	77.97 ± 3.06	<.001[†]
		non-csPC	11.08 ± 1.85	80.12 ± 2.71	<.001[†]

↑ larger better   DSC = Dice Score Coefficient, AUC = Area under the curve.
[†] Reference is idealized scenario. P<0.05 were considered statistically significant.

Previous studies in the literature have tackled the csPC detection from MRI with compound systems, yielding to AUCs in the range of 0.81–0.91 [2,18,21]. However, the evaluation of the system is presented in terms of the detection ability as a standalone module and without accounting for faults stemming in the preceding segmentation modules. In contract to those studies, we aim at characterizing emergent behaviors stemming from the interaction of the modules in the compound PC system. Despite not being directly comparable, our results suggest significant drops in csPC detection performance when accounting for the interaction between modules.

With the expected increase in AI developments based on MRI for PC [8], our study highlights the importance of a holistic evaluation for compound PC

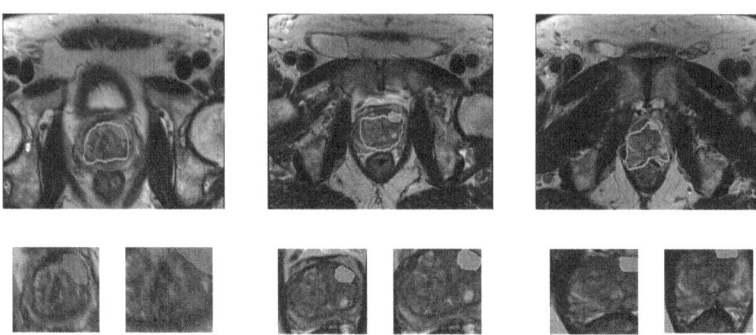

Fig. 3. Qualitative results from the segmentation module and in the presence of two different segmentation architectures. In yellow, results for nnU-Net. In red, results for tU-Net. Bottom row depicts the eventual input for the detection module, based on the output of the segmentation one. (Color figure online)

systems. In that regard, our results suggest that the lack of exhaustive evaluations can lead to overoptimistic evaluation reports. Hereby, a comprehensive independent evaluation of each module and of the interaction between modules could benefit the translation of the systems into clinical practice [6,24].

Limitations. First, our study was limited by its retrospective nature and lack of external evaluation. Second, our study was limited to compound PC systems consisting of a segmentation module and a detection module. In spite of their wide adoption, other configurations should be considered in future studies. Finally, we restricted our study to U-Net based architectures. Future studies considering other widely adopted architectures for PC detection might provide additional insights about the interactions effect in compound PC systems.

5 Conclusion

In conclusion, our results indicate that unaccounted interactions between modules of compound csPC detection systems might lead to overoptimistic evaluation results. We found that when accounting for the interaction between modules, undesired behaviors emerge leading to significant drops in performance. To avoid it, we suggest that evaluations of PC compound systems should be holistic, presenting the performance of independent modules and when interacting. Our observations could potentially increase the adoption of AI developments for PC and MRI in clinical practice.

Acknowledgements. The authors would like to show their gratitude to the organizers of the ProstateX challenge for providing access to their curated dataset.

Disclosure of Interests. The authors declare that they have no competing interests.

References

1. Anaya-Isaza, A., Mera-Jiménez, L., Fernandez-Quilez, A.: Crosstransunet: a new computationally inexpensive tumor segmentation model for brain mri. IEEE Access **11**, 27066–27085 (2023)
2. Cao, R., et al.: Joint prostate cancer detection and Gleason score prediction in MP-MRI via focalnet. IEEE Trans. Med. Imaging **38**(11), 2496–2506 (2019)
3. Cuocolo, R., et al.: Deep learning whole-gland and zonal prostate segmentation on a public MRI dataset. J. Magn. Reson. Imaging **54**(2), 452–459 (2021)
4. Eklund, M., et al.: MRI-targeted or standard biopsy in prostate cancer screening. N. Engl. J. Med. **385**(10), 908–920 (2021)
5. Fassia, M.K., et al.: Deep learning prostate MRI segmentation accuracy and robustness: a systematic review. Radiol. Artif. Intell. **2014**, e230138 (2024)
6. Fernandez-Quilez, A.: Deep learning in radiology: ethics of data and on the value of algorithm transparency, interpretability and explainability. AI Ethics **3**(1), 257–265 (2023)
7. Fernandez-Quilez, A., Larsen, S.V., Goodwin, M., Gulsrud, T.O., Kjosavik, S.R., Oppedal, K.: Improving prostate whole gland segmentation in t2-weighted MRI with synthetically generated data. In: 2021 IEEE 18th International Symposium on Biomedical Imaging (ISBI), pp. 1915–1919. IEEE (2021)
8. Fernandez-Quilez, A., Nordström, T., Jäderling, F., Kjosavik, S.R., Eklund, M.: Prostate age gap: an MRI surrogate marker of aging for prostate cancer detection. J. Magn. Reson. Imaging **60**(2), 458–468 (2024). https://doi.org/10.1002/jmri.29090
9. Isensee, F., Jaeger, P.F., Kohl, S.A., Petersen, J., Maier-Hein, K.H.: NNU-net: a self-configuring method for deep learning-based biomedical image segmentation. Nat. Methods **18**(2), 203–211 (2021)
10. Kurbatskaya, A., Jaramillo-Jimenez, A., Ochoa-Gomez, J.F., Brnnick, K., Fernandez-Quilez, A.: Assessing gender fairness in EEG-based machine learning detection of Parkinson's disease: a multi-center study. In: 2023 31st European Signal Processing Conference (EUSIPCO), pp. 1020–1024 (2023). https://doi.org/10.23919/EUSIPCO58844.2023.10289837
11. Kwee, T.C., Kwee, R.M.: Workload of diagnostic radiologists in the foreseeable future based on recent scientific advances: growth expectations and role of artificial intelligence. Insights Imaging **12**(1), 1–12 (2021)
12. van Leeuwen, K.G., Schalekamp, S., Rutten, M.J., van Ginneken, B., de Rooij, M.: Artificial intelligence in radiology: 100 commercially available products and their scientific evidence. Eur. Radiol. **31**, 3797–3804 (2021)
13. Lindeijer, T.N., et al.: Leveraging multi-view data without annotations for prostate MRI segmentation: a contrastive approach. arXiv preprint arXiv:2308.06477 (2023)
14. Litjens, G., Debats, O., Barentsz, J., Karssemeijer, N., Huisman, H.: Computer-aided detection of prostate cancer in MRI. IEEE Trans. Med. Imaging **33**(5), 1083–1092 (2014)
15. Litjens, G., et al.: Evaluation of prostate segmentation algorithms for MRI: the promise12 challenge. Med. Image Anal. **18**(2), 359–373 (2014)
16. McKinney, S.M., et al.: International evaluation of an AI system for breast cancer screening. Nature **577**(7788), 89–94 (2020)
17. Ronneberger, O., Fischer, P., Brox, T.: U-net: convolutional networks for biomedical image segmentation. In: Navab, N., Hornegger, J., Wells, W.M., Frangi, A.F. (eds.) MICCAI 2015. LNCS, vol. 9351, pp. 234–241. Springer, Cham (2015). https://doi.org/10.1007/978-3-319-24574-4_28

18. Saha, A., et al.: Artificial intelligence and radiologists in prostate cancer detection on MRI (PI-CAI): an international, paired, non-inferiority, confirmatory study. Lancet Oncol. (2024)
19. Saha, A., Hosseinzadeh, M., Huisman, H.: End-to-end prostate cancer detection in BPMRI via 3D CNNs: effects of attention mechanisms, clinical priori and decoupled false positive reduction. Med. Image Anal. **73**, 102155 (2021)
20. Sanford, T., et al.: Deep-learning-based artificial intelligence for PI-RADS classification to assist multiparametric prostate MRI interpretation: a development study. J. Magn. Reson. Imaging **52**(5), 1499–1507 (2020)
21. Sanyal, J., Banerjee, I., Hahn, L., Rubin, D.: An automated two-step pipeline for aggressive prostate lesion detection from multi-parametric MR sequence. AMIA Summits Transl. Sci. Proc. **2020**, 552 (2020)
22. Suarez-Ibarrola, R., et al.: Artificial intelligence in magnetic resonance imaging-based prostate cancer diagnosis: where do we stand in 2021? Eur. Urol. Focus **8**(2), 409–417 (2022)
23. Wenderott, K., Krups, J., Luetkens, J.A., Gambashidze, N., Weigl, M.: Prospective effects of an artificial intelligence-based computer-aided detection system for prostate imaging on routine workflow and radiologists' outcomes. Eur. J. Radiol. **170**, 111252 (2024)
24. Widner, K., et al.: Lessons learned from translating AI from development to deployment in healthcare. Nat. Med. **29**(6), 1304–1306 (2023)
25. Yushkevich, P.A., Gao, Y., Gerig, G.: ITK-SNAP: an interactive tool for semi-automatic segmentation of multi-modality biomedical images. In: 2016 38th Annual International Conference of the IEEE Engineering in Medicine and Biology Society (EMBC), pp. 3342–3345. IEEE (2016)

Optimizing Multi-expert Consensus for Classification and Precise Localization of Barrett's Neoplasia

Carolus H. J. Kusters[1]( ), Tim G. W. Boers[1], Tim J. M. Jaspers[1], Martijn R. Jong[2], Rixta A. H. van Eijck van Heslinga[2], Albert J. de Groof[2], Jacques J. Bergman[2], Fons van der Sommen[1], and Peter H. N. De With[1]

[1] Department of Electrical Engineering, Video Coding and Architectures, Eindhoven University of Technology, Eindhoven, The Netherlands
{c.h.j.kusters,t.boers,t.j.m.jaspers,fvdsommen,p.h.n.de.with}@tue.nl
[2] Department of Gastroenterology and Hepatology, Amsterdam University Medical Centers, University of Amsterdam, Amsterdam, The Netherlands

Abstract. Recognition of early neoplasia in Barrett's Esophagus (BE) is challenging, despite advances in endoscopic technology. Even with correct identification, the subtle nature of lesions leads to significant inter-observer variability in placing targeted biopsy markers and delineation of lesions. Computer-Aided Detection (CADe) systems may assist endoscopists, however, compliance of endoscopists with CADe is often suboptimal, reducing joint performance below CADe stand-alone performance. Improved localization performance of CADe could enhance compliance. These systems often use fused consensus ground-truths (GT), which may not capture subtle neoplasia gradations, affecting classification and localization. This study evaluates five consensus GT strategies from multi-expert segmentation labels and four loss functions for their impact on classification and localization performance. The dataset includes 7,995 non-dysplastic BE images (1,256 patients) and 2,947 neoplastic images (823 patients), with each neoplastic image annotated by two experts. Classification, localization for true positives, and combined detection performance are assessed and compared with 14 independent Barrett's experts. Results show that using multiple consensus GT masks with a compound Binary Cross-Entropy and Dice loss achieves the best classification sensitivity and near-expert level localization, making it the most effective training strategy. The code is made publicly available at: https://github.com/BONS-AI-VCA-AMC/BE-CADe-GT.

Keywords: Barrett's Esophagus · Detection · Inter-observer variability

1 Introduction

Barrett's Esophagus (BE) is a well-known precursor for Esophageal Adenocarcinoma (EAC). BE patients undergo regular endoscopic surveillance to enable early diagnosis and treatment with a good prognosis [15,19]. However, detection of early BE neoplasia is challenging due to the subtle endoscopic appearance and sampling errors from extensive random biopsy protocols [3]. Although recent endoscopes allow visualization of nearly all of such subtle changes, recognition of neoplasia by endoscopists is the limiting factor. Despite correct identification, the subtle nature of lesions often causes endoscopists to struggle with optimal biopsy placement for pathological confirmation, potentially leading to disease understaging or missed neoplasia. Furthermore, it leads to significant inter-observer variability among Barrett's experts, reflected in inconsistent delineation of neoplasia [2], a challenge also common to other applications [12,14].

Several Computer-Aided Detection (CADe) systems for Barrett's neoplasia have been developed to assist endoscopists in the recognition of neoplastic lesions [1,5,6,8,10,13]. However, endoscopist compliance with CADe if often suboptimal, leading to joint sensitivity being lower than CADe alone [6]. These systems focus on the detection (i.e. binary classification and localization) of neoplastic lesions, requiring pixel-level annotations by Barrett's experts, in which inter-observer variability can degrade performance of models trained with suboptimal segmentation labels [20]. To mitigate this, a fused consensus ground-truth from multiple experts is often used [1,6,8]. However, these binary masks may fail to capture the subtle gradations of neoplasia with a broad range of atypical patterns, crucial for detecting particularly subtle variants. Missing these nuances can prevent the model from accurately identifying the most abnormal tissue area for targeted biopsy, leading to understaging or missed neoplasia. We hypothesize that the compliance of endoscopists may be increased by improved localization performance of CADe, as localizations are often suboptimal or even incorrect.

De Groof *et al.* [8] and Hussein *et al.* [10] evaluated the localization performance of predicted targeted biopsy sites against various consensus masks, but did not explore the impact of inter-observer variability and fusion techniques for consensus ground-truths nor the loss function choice for model training. In contrast, Van der Sommen *et al.* [17] and Boers *et al.* [2] investigated these aspects, but focused solely on primary detection, without assessing localization.

This extends on the investigation of Van der Sommen *et al.* [17] and Boers *et al.* [2], by proposing and investigating several consensus ground-truths from multi-expert segmentation labels and loss functions for training segmentation models, to optimize both binary classification and localization performance. While the primary focus is on Barrett's neoplasia detection, we expect the conceptual solution to be applicable to other medical fields as well.

2 Methods and Materials

2.1 Data

A. Setting: This study utilizes private internal data collected both retrospectively and prospectively from 15 international centers participating in the BONS-AI consortium (Barrett's OesophaguS imaging for Artificial Intelligence) consortium. All data, acquired using Olympus gastroscopes (Olympus Corp.,

Tokyo, Japan), includes Barrett's neoplasia and non-dysplastic BE (NDBE) with histopathology confirmation by means of biopsy or resection. Data are curated to indicate the presence or absence of a visible lesion for neoplasia and NDBE, respectively. The detailed data acquisition protocol is published elsewhere [7].

B. Datasets: The training set consists of 2,747 neoplastic images (715 patients) and 7,595 NDBE images (1,095 patients). The validation set used for model checkpoint selection consists of 100 neoplastic images (58 patients) and 100 NDBE images (36 patients), while the test set consists of 100 neoplastic images (50 patients) and 300 NDBE images (125 patients). The dataset split is made in consultation with clinicians to ensure representative training and validation sets, while the test set is enriched with challenging subtle neoplastic cases. A strict patient-based split is employed to avoid data leakage and intra-patient bias. Neoplastic delineations are obtained from 14 BONS-AI Barrett's experts, where at least 2 experts delineated the same image. Experts outlined the largest area atypical from normal NDBE, suspected to be neoplasia (Lower Likelihood, LL), and the area within it that stands out more profoundly (Higher Likelihood, HL). If the HL delineations obtained $\leq 30\%$ agreement by Dice Score, a third expert was consulted. The two delineations with the highest overlap were used for further ground-truth (GT) processing to achieve a minimum consensus level.

C. Segmentation Ground-Truth (GT): Each neoplastic image has four delineations: a lower likelihood annotation (LL) and a higher likelihood annotation (HL), independently provided by two experts. Several consensus GT masks with increasing neoplastic certainty can be constructed: Soft spot GT, Plausible spot GT, Sweet spot GT and Hard spot GT, which are defined in Table 1. For model training, also Average GT, Random GT and Multiple GT strategies are considered, also in Table 1. An example case with GT masks is illustrated in Fig. 1.

Table 1. Definitions for segmentation ground-truth strategies. Subscript indices for the LL and HL delineations correspond to the respective expert numbers.

Consensus GT	Definition	Description of delineations
Soft spot	$LL_1 \cup LL_2$	All neoplastic and/or atypical
Plausible spot	$(LL_1 \cap LL_2) \cup HL_1 \cup HL_2$	Neoplastic by ≥ 1 expert and atypical by both experts
Sweet spot	$HL_1 \cup HL_2$	Neoplastic by ≥ 1 expert
Hard spot	$HL_1 \cap HL_2$	Neoplastic by both experts
Average	Averaged consensus masks in single non-binary mask	
Random	Randomly sampling single consensus mask during training	
Multiple	Provide all consensus masks during training	

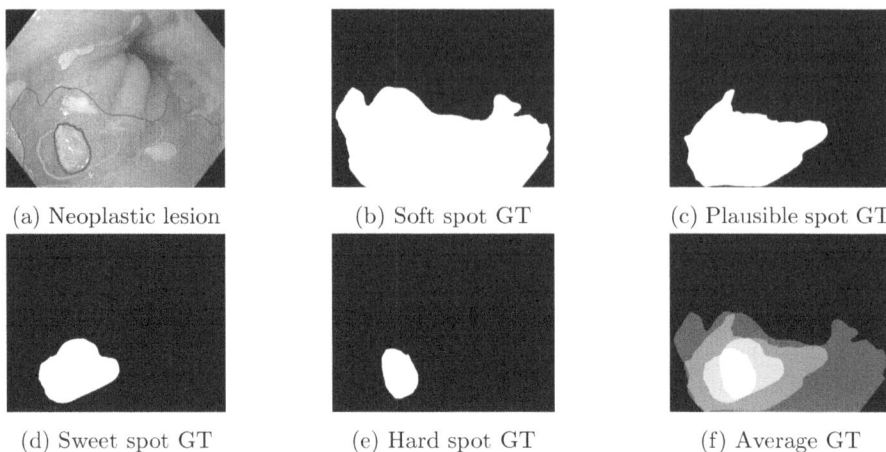

Fig. 1. Example lesion with delineations and consensus GTs. (a) Neoplastic lesion; thick green/blue lines indicate HL delineations, and thinner lines indicate LL delineations. (b)–(e) Corresponding consensus GT masks. (f) Consensus GT masks pooled into non-binary average map. For Random GT one of (b)–(e) is sampled randomly, while for the Multiple GT (b)–(e) are all provided. (Color figure online)

D. Expert Benchmark: To establish a performance reference, 14 internationally recognized Barrett's experts, not affiliated with BONS-AI and encountering all cases for the first time, evaluated the test set. Each expert reviewed 50 neoplastic images (50 patients) and 150 randomly selected NDBE images. Experts assessed for the presence of neoplasia (binary classification) and placed a targeted biopsy marker on the most abnormal area of the detected lesion, resulting in 2 to 8 biopsy markers per neoplastic image. An example case with expert biopsy markers (green dots) is illustrated in Fig. 2. The classification, localization for True Positive (TP) classifications, and combined detection expert performances are summarized in Table 2. Further details are provided in our previous work [6].

2.2 Training and Evaluation Details

A. Network Architecture: The U-Net [16] architecture with a ResNet-50 [9] backbone, well-established for medical image segmentation, is chosen as the model. The backbone is initialized with ImageNet [4] pre-trained weights and a subsequent classification head, comprising an average pooling and a fully-connected layer with Sigmoid activation, is attached after the final feature extraction layer. The model is implemented with PyTorch segmentation models [11].

B. Data Pre-processing: The active region of raw endoscopic images is resized to 256×256 pixels and normalized channel-wise with the mean and standard deviation of training data. Data augmentation techniques are then applied, including random combinations of horizontal and vertical flipping, rotation by $\theta \in \{0°, 90°, 180°, 270°\}$, contrast/saturation/brightness changes, gray-scale

Table 2. Performance of 14 Barrett's Experts on classification, localization and the combined detection on the test set. Results are presented as Mean ± Std.

Objective	Acc	Sens	Spec
Classification	0.847 ± 0.077	0.851 ± 0.097	0.845 ± 0.121
Localization (Soft GT)	0.962 ± 0.027	N/A	N/A
Localization (Plausible GT)	0.938 ± 0.046	N/A	N/A
Localization (Sweet GT)	0.891 ± 0.044	N/A	N/A
Localization (Hard GT)	0.757 ± 0.062	N/A	N/A
Detection (Soft GT)	0.839 ± 0.079	0.819 ± 0.088	0.845 ± 0.121
Detection (Plausible GT)	0.834 ± 0.081	0.799 ± 0.094	0.845 ± 0.121
Detection (Sweet GT)	0.824 ± 0.083	0.759 ± 0.094	0.845 ± 0.121
Detection (Hard GT)	0.795 ± 0.085	0.643 ± 0.078	0.845 ± 0.121

conversion, Gaussian blurring, random affine and sharpness transforms, and random Gaussian noise corruption. To obtain a reliable performance estimate, validation is performed four times on different randomly augmented validation sets, increasing the sample size by a factor four, using the first three techniques.

C. Training Procedures: All experiments are trained with batch sizes of 32 and a learning rate of 10^{-4} for 150 epochs, using the Adam optimizer with AMSgrad, $(\beta_1, \beta_2)=(0.9, 0.999)$ and a weight decay of 10^{-4}. A learning-rate scheduler reduces the learning rate by a factor of 10 after 10 epochs without validation loss improvement, up to three reductions. Early stopping occurs if the validation classification metric does not improve for 25 epochs. Binary Cross-Entropy (BCE) loss is used for classification, while various segmentation losses with different GT strategies are evaluated as outlined in Sects. 2.1-C and 2.2-D. Randomly sampling training images ensures approximately 50–50% class representation in each iteration. Each experiment is performed three times with different random seeds and for each experiment the checkpoint with the best validation classification performance is used. Experiments are implemented in Python 3.10 with PyTorch (Lightning) and executed on a A100 GPU (NVIDIA Corp., CA, USA).

D. Segmentation Loss Functions: The architecture is trained using four different segmentation loss functions, namely Mean-Squared Error (MSE), Binary Cross-Entropy (BCE), Dice and a compound BCE and Dice loss (BCE+Dice). For the average GT mask training strategy, only MSE and BCE are evaluated, as Dice is only suitable for binary GT masks. For the multiple GT mask training strategy, the loss functions are extended to compute the average loss between the predicted segmentation mask and the four individual consensus GT masks.

E. Performance Evaluation: The model's performance is evaluated on classification, localization, and combined detection, similar to the assessment of Barrett's experts. Classification performance is measured using the Area Under the Receiver Operating Characteristic Curve (AUROC). Accuracy, sensitivity and specificity are determined by selecting a threshold to achieve $\geq 90\%$ sensitivity on the validation set. Localization performance for True Positive (TP) classifications is assessed by checking if the predicted biopsy site, indicated by the maximal pixel value in the segmentation map within a 2-pixel radius, overlaps with the consensus GT masks. A TP for combined detection requires both correct classification and localization; otherwise, it is considered a False Negative.

3 Experimental Results and Discussion

The classification, localization and combined detection performances for experts and the obtained results for all experiments are presented in Tables 2 and 3, respectively. Figure 2 provides examples with visualizations of the predicted segmentation maps, expert and predicted biopsy sites, and respective GT masks.

1. Expert Performance: The international Barrett's experts achieved accuracy, sensitivity and specificity of 84.7%, 85.1% and 84.5% on the classification task, respectively. Localization performance ranges from 96.2% for the Soft GT to 75.7% for the Hard GT consensus masks, highlighting challenges in accurately placing biopsy markers in the most certain neoplastic area of expert annotations. However, this variation may be largely described by inter-observer variability in delineations as a given uncertainty aspect. Incorrect localization significantly impacts combined detection sensitivity, ranging between 81.9% and 64.3%.

2. Model Performance: The results demonstrate that using the Multiple GT strategy with the BCE+Dice loss function achieves the best classification performance, with an AUROC of 94.2%. The Multiple GT strategy with BCE and BCE+Dice shows the best accuracy and sensitivity, respectively, while Soft GT with MSE excels in specificity. Training with multiple masks improves sensitivity and AUROC by incorporating unweighted supervision with masks containing various levels of neoplastic certainty during loss calculation.

In terms of localization performance, models using Hard spot GT with Dice and BCE+Dice loss functions outperform others across all GT masks, with performance ranging from 96.5% to 74.9%. However, this performance may be influenced by lower sensitivity in the primary classification task, as harder-to-locate lesions are missed and excluded from localization evaluation, thus artificially boosting localization metrics. The results suggest that except for the Soft spot GT strategy, the expert localization performance can be approached.

The predicted segmentation masks in Fig. 2 show that training with Soft spot GT results in large confident predictions, Hard spot GT results in very condensed predictions and Random GT falls in between. Average GT and Multiple GT masks improve coverage of subtle neoplasia gradations, delineating the extent of atypical tissue while identifying the most abnormal regions for targeted biopsy.

For the combined detection performance, the Multiple GT strategy with BCE+Dice loss achieves the highest sensitivity, ranging from 85.3% to 65.0%. Combining excellent classification sensitivity and near-expert level localization makes it the most effective training strategy, as sensitivity is prioritized in CADe systems for BE neoplasia given the higher cost of FNs compared to FPs [6].

The obtained results underscore the benefit of incorporating gradation into segmentation ground-truth, enabling the model to more effectively learn nuanced neoplasia characteristics in BE. The significant performance differences between strategies, suggest that the results may translate to other fields.

Table 3. Classification, localization and combined detection performance on the test set. Results are presented as Mean ± Std. The best and second best results are highlighted in boldface and underlined, respectively.

Training Strategy		Classification				Localization			
GT	Loss	Acc	Sens	Spec	AUROC	Acc (Soft)	Acc (Plausible)	Acc (Sweet)	Sens (Hard)
Soft	MSE	0.872 ± 0.029	0.837 ± 0.042	**0.883 ± 0.036**	0.922 ± 0.008	0.886 ± 0.069	0.847 ± 0.086	0.791 ± 0.066	0.627 ± 0.066
Soft	BCE	0.788 ± 0.127	0.867 ± 0.035	0.762 ± 0.175	0.902 ± 0.031	0.847 ± 0.049	0.789 ± 0.046	0.739 ± 0.083	0.550 ± 0.068
Soft	Dice	0.866 ± 0.015	0.823 ± 0.042	0.880 ± 0.033	0.923 ± 0.012	0.742 ± 0.095	0.652 ± 0.132	0.499 ± 0.134	0.337 ± 0.105
Soft	BCE+Dice	0.795 ± 0.093	0.860 ± 0.044	0.773 ± 0.137	0.915 ± 0.020	0.909 ± 0.064	0.886 ± 0.073	0.836 ± 0.073	0.605 ± 0.141
Hard	MSE	0.827 ± 0.049	0.840 ± 0.062	0.822 ± 0.086	0.910 ± 0.006	0.877 ± 0.061	0.874 ± 0.055	0.846 ± 0.060	0.725 ± 0.074
Hard	BCE	0.858 ± 0.011	0.850 ± 0.035	0.860 ± 0.006	0.923 ± 0.013	0.932 ± 0.038	0.921 ± 0.033	0.889 ± 0.034	0.739 ± 0.063
Hard	Dice	0.864 ± 0.008	0.863 ± 0.012	0.864 ± 0.013	0.924 ± 0.006	0.946 ± 0.027	**0.938 ± 0.034**	**0.911 ± 0.037**	**0.749 ± 0.019**
Hard	BCE+Dice	0.855 ± 0.035	0.850 ± 0.017	0.857 ± 0.049	0.922 ± 0.004	**0.965 ± 0.035**	**0.938 ± 0.040**	0.906 ± 0.022	**0.749 ± 0.032**
Random	MSE	0.835 ± 0.028	0.880 ± 0.017	0.820 ± 0.044	0.920 ± 0.014	0.917 ± 0.012	0.890 ± 0.009	0.845 ± 0.006	0.682 ± 0.013
Random	BCE	0.820 ± 0.059	0.823 ± 0.084	0.819 ± 0.095	0.906 ± 0.027	0.934 ± 0.033	0.917 ± 0.041	0.876 ± 0.042	0.737 ± 0.008
Random	Dice	0.827 ± 0.060	0.853 ± 0.025	0.818 ± 0.088	0.918 ± 0.012	0.911 ± 0.027	0.891 ± 0.004	0.816 ± 0.035	0.570 ± 0.015
Random	BCE+Dice	0.796 ± 0.048	0.857 ± 0.025	0.776 ± 0.072	0.906 ± 0.019	0.923 ± 0.031	0.887 ± 0.040	0.857 ± 0.025	0.717 ± 0.056
Average	MSE	0.833 ± 0.034	0.840 ± 0.044	0.830 ± 0.045	0.908 ± 0.012	0.940 ± 0.012	0.912 ± 0.017	0.893 ± 0.024	0.719 ± 0.031
Average	BCE	0.831 ± 0.027	0.803 ± 0.012	0.840 ± 0.038	0.888 ± 0.012	0.896 ± 0.015	0.871 ± 0.043	0.830 ± 0.042	0.697 ± 0.042
Multiple	MSE	0.841 ± 0.048	0.850 ± 0.040	0.839 ± 0.074	0.918 ± 0.006	0.934 ± 0.015	0.887 ± 0.019	0.855 ± 0.011	0.679 ± 0.039
Multiple	BCE	**0.873 ± 0.031**	0.887 ± 0.021	0.868 ± 0.036	0.922 ± 0.010	0.925 ± 0.019	0.891 ± 0.030	0.841 ± 0.054	0.691 ± 0.080
Multiple	Dice	0.853 ± 0.019	0.883 ± 0.006	0.843 ± 0.023	0.931 ± 0.008	0.891 ± 0.028	0.842 ± 0.040	0.736 ± 0.054	0.510 ± 0.055
Multiple	BCE+Dice	0.845 ± 0.060	**0.900 ± 0.026**	0.827 ± 0.085	**0.942 ± 0.004**	0.948 ± 0.022	0.923 ± 0.027	0.878 ± 0.058	0.722 ± 0.015

		Detection							
		Acc (Soft)	Sens (Soft)	Acc (Plausible)	Sens (Plausible)	Acc (Sweet)	Sens (Sweet)	Acc (Hard)	Sens (Hard)
Soft	MSE	0.848 ± 0.038	0.740 ± 0.046	0.839 ± 0.039	0.707 ± 0.049	0.828 ± 0.035	0.660 ± 0.035	0.793 ± 0.036	0.523 ± 0.038
Soft	BCE	0.755 ± 0.136	0.733 ± 0.021	0.743 ± 0.137	0.683 ± 0.045	0.732 ± 0.145	0.640 ± 0.072	0.691 ± 0.143	0.477 ± 0.061
Soft	Dice	0.813 ± 0.025	0.610 ± 0.075	0.704 ± 0.020	0.537 ± 0.106	0.763 ± 0.033	0.410 ± 0.106	0.720 ± 0.028	0.277 ± 0.083
Soft	BCE+Dice	0.775 ± 0.107	0.780 ± 0.017	0.770 ± 0.110	0.760 ± 0.026	0.759 ± 0.107	0.717 ± 0.031	0.709 ± 0.115	0.517 ± 0.100
Hard	MSE	0.801 ± 0.054	0.737 ± 0.070	0.800 ± 0.054	0.733 ± 0.065	0.794 ± 0.055	0.710 ± 0.066	0.768 ± 0.066	0.607 ± 0.040
Hard	BCE	0.843 ± 0.018	0.793 ± 0.064	0.841 ± 0.018	0.783 ± 0.057	0.834 ± 0.018	0.757 ± 0.059	0.803 ± 0.022	0.630 ± 0.078
Hard	Dice	0.853 ± 0.011	0.817 ± 0.031	**0.851 ± 0.013**	0.810 ± 0.036	**0.845 ± 0.009**	0.787 ± 0.042	**0.810 ± 0.011**	0.647 ± 0.021
Hard	BCE+Dice	0.848 ± 0.035	0.820 ± 0.017	0.842 ± 0.035	0.797 ± 0.021	0.835 ± 0.035	0.770 ± 0.010	0.802 ± 0.043	0.637 ± 0.025
Random	MSE	0.817 ± 0.030	0.807 ± 0.015	0.811 ± 0.027	0.783 ± 0.023	0.801 ± 0.029	0.743 ± 0.015	0.765 ± 0.033	0.600 ± 0.000
Random	BCE	0.807 ± 0.053	0.770 ± 0.098	0.803 ± 0.052	0.757 ± 0.103	0.795 ± 0.055	0.723 ± 0.110	0.766 ± 0.062	0.607 ± 0.055
Random	Dice	0.808 ± 0.066	0.777 ± 0.006	0.803 ± 0.061	0.760 ± 0.020	0.788 ± 0.063	0.697 ± 0.035	0.735 ± 0.063	0.487 ± 0.021
Random	BCE+Dice	0.779 ± 0.056	0.790 ± 0.017	0.772 ± 0.056	0.760 ± 0.036	0.765 ± 0.054	0.733 ± 0.006	0.735 ± 0.061	0.613 ± 0.031
Average	MSE	0.820 ± 0.031	0.790 ± 0.046	0.814 ± 0.032	0.767 ± 0.051	0.810 ± 0.029	0.750 ± 0.050	0.773 ± 0.029	0.603 ± 0.023
Average	BCE	0.810 ± 0.025	0.720 ± 0.020	0.805 ± 0.020	0.700 ± 0.036	0.797 ± 0.020	0.667 ± 0.040	0.770 ± 0.021	0.560 ± 0.040
Multiple	MSE	0.828 ± 0.050	0.793 ± 0.025	0.818 ± 0.052	0.753 ± 0.021	0.811 ± 0.051	0.727 ± 0.025	0.773 ± 0.057	0.577 ± 0.031
Multiple	BCE	**0.856 ± 0.035**	0.820 ± 0.035	0.848 ± 0.037	0.790 ± 0.044	0.838 ± 0.043	0.747 ± 0.064	0.804 ± 0.046	0.613 ± 0.080
Multiple	Dice	0.829 ± 0.015	0.787 ± 0.021	0.818 ± 0.011	0.743 ± 0.031	0.795 ± 0.008	0.650 ± 0.044	0.745 ± 0.007	0.450 ± 0.046
Multiple	BCE+Dice	0.833 ± 0.060	**0.853 ± 0.015**	0.828 ± 0.060	**0.830 ± 0.017**	0.818 ± 0.061	**0.790 ± 0.017**	0.783 ± 0.059	**0.650 ± 0.020**

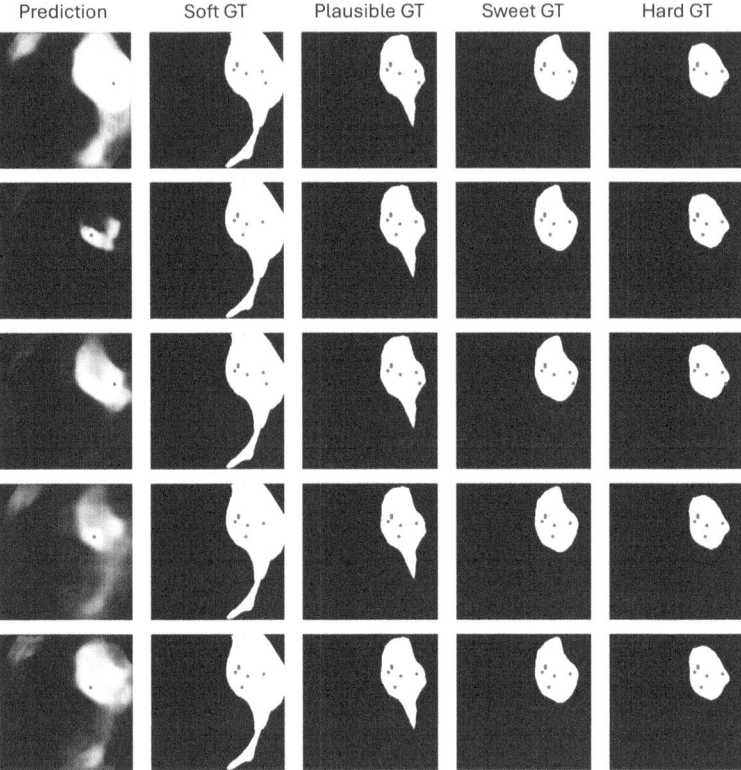

Fig. 2. Example case with visualizations of the predicted segmentation maps, predicted biopsy site (red dot), and expert biopsy sites (green dots) alongside the respective GT masks. From top to bottom models trained with: Soft spot GT, Hard spot GT, Random GT, Average GT and Multiple GT. (Color figure online)

4 Conclusions

We have evaluated various training strategies using consensus GT from multi-expert segmentation labels and different loss functions for detecting Barrett's neoplasia, focusing on both classification and localization. Our results demonstrate that the choice of segmentation GT significantly impacts model performance regarding primary classification, localization and gradual neoplasia coverage of predicted segmentation masks. Therefore, it is crucial to consider these effects and incorporate measures to create awareness of uncertainty in the data. Future work should explore some alternatives. First, the predicted targeted biopsy site was determined by maximal pixel value, however, the geometric center of the prediction could also be viable [10]. Second, although this study employed a single well-established segmentation network architecture, as the focus was not on the architecture itself, future experiments could explore various state-of-the-art segmentation architectures and backbones. Lastly, exclusively deterministic models and training strategies are assessed, whereas probabilistic models may better capture uncertainty from multi-annotated data [18].

Acknowledgements. This work is facilitated by data/equipment from Olympus Corp., Tokyo, Japan. We gratefully acknowledge their provided research support. We thank SURF (www.surf.nl) for the support in using the National Supercomputer Snellius.

Disclosure of Interests. Authors have no conflict of interest to declare.

References

1. Abdelrahim, M., et al.: Development and validation of artificial neural networks model for detection of Barrett's neoplasia: a multicenter pragmatic nonrandomized trial (with video). Gastrointest. Endosc. **97**(3), 422–434 (2023). https://doi.org/10.1016/j.gie.2022.10.031
2. Boers, T.G.W., et al.: Comparing training strategies using multi-assessor segmentation labels for-Barrett's neoplasia detection. In: Ali, S., van der Sommen, F., Papież, B.W., van Eijnatten, M., Jin, Y., Kolenbrander, I. (eds.) Cancer Prevention Through Early Detection, pp. 131–138. Springer, Cham (2022). https://doi.org/10.1007/978-3-031-17979-2_13
3. Davis-Yadley, A.H., Neill, K.G., Malafa, M.P., Peña, L.R.: Advances in the endoscopic diagnosis of Barrett esophagus. Cancer Control: J. Moffitt Cancer Center **23**(1), 67–77 (2016). https://api.semanticscholar.org/CorpusID:24893806
4. Deng, J., Dong, W., Socher, R., Li, L.J., Li, K., Fei-Fei, L.: Imagenet: a large-scale hierarchical image database. In: 2009 IEEE Conference on Computer Vision and Pattern Recognition, pp. 248–255 (2009).https://doi.org/10.1109/CVPR.2009.5206848
5. Ebigbo, A., et al.: Real-time use of artificial intelligence in the evaluation of cancer in Barrett's oesophagus. Gut **69**(4), 615–616 (2020)
6. Fockens, K.N., et al.: A deep learning system for detection of early Barrett's neoplasia: a model development and validation study. Lancet Digit. Health **5**(12), e905–e916 (2023). https://doi.org/10.1016/S2589-7500(23)00199-1
7. Fockens, K.N., et al.: Towards a robust and compact deep learning system for primary detection of early Barrett's neoplasia: initial image-based results of training on a multi-center retrospectively collected data set. United Eur. Gastroenterol. J. **11**(4), 324–336 (2023)
8. de Groof, A.J., et al.: Deep-learning system detects neoplasia in patients with Barrett's esophagus with higher accuracy than endoscopists in a multistep training and validation study with benchmarking. Gastroenterology **158**(4), 915–929 (2020)
9. He, K., Zhang, X., Ren, S., Sun, J.: Deep residual learning for image recognition. In: 2016 IEEE Conference on Computer Vision and Pattern Recognition (CVPR), pp. 770–778 (2016). https://doi.org/10.1109/CVPR.2016.90
10. Hussein, M., et al.: A new artificial intelligence system successfully detects and localises early neoplasia in Barrett's esophagus by using convolutional neural networks. United Eur. Gastroenterol. J. **10**(6), 528–537 (2022). https://doi.org/10.1002/ueg2.12233
11. Iakubovskii, P.: Segmentation Models Pytorch (2019). https://github.com/qubvel/segmentation_models.pytorch

12. Joskowicz, L., Cohen, D., Caplan, N., Sosna, J.: Inter-observer variability of manual contour delineation of structures in CT. Eur. Radiol. **29** (2018). https://doi.org/10.1007/s00330-018-5695-5
13. Meinikheim, M., et al.: Influence of artificial intelligence on the diagnostic performance of endoscopists in the assessment of Barrett's esophagus: a tandem randomized and video trial. Endoscopy **56**(09), 641–649 (2024). https://doi.org/10.1055/a-2296-5696
14. Menze, B.H., et al.: The multimodal brain tumor image segmentation benchmark (Brats). IEEE Trans. Med. Imaging **34**(10), 1993–2024 (2015). https://doi.org/10.1109/TMI.2014.2377694
15. Pech, O., et al.: Long-term efficacy and safety of endoscopic resection for patients with mucosal adenocarcinoma of the esophagus. Gastroenterology **146**(3), 652–660 (2014)
16. Ronneberger, O., Fischer, P., Brox, T.: U-Net: convolutional networks for biomedical image segmentation. In: Navab, N., Hornegger, J., Wells, W.M., Frangi, A.F. (eds.) MICCAI 2015. LNCS, vol. 9351, pp. 234–241. Springer, Cham (2015). https://doi.org/10.1007/978-3-319-24574-4_28
17. van der Sommen, F., Zinger, S., Schoon, E.J., de With, P.H.N.: Sweet-spot training for early esophageal cancer detection. In: Tourassi, G.D., III, S.G.A. (eds.) Medical Imaging 2016: Computer-Aided Diagnosis, vol. 9785, p. 97851B. International Society for Optics and Photonics, SPIE (2016). https://doi.org/10.1117/12.2208114
18. Valiuddin, M.M.A., Viviers, C.G.A., van Sloun, R.J.G., de With, P.H.N., van der Sommen, F.: Improving aleatoric uncertainty quantification in multi-annotated medical image segmentation with normalizing flows. In: Sudre, C.H., et al. (eds.) UNSURE/PIPPI -2021. LNCS, vol. 12959, pp. 75–88. Springer, Cham (2021). https://doi.org/10.1007/978-3-030-87735-4_8
19. Weusten, B., et al.: Endoscopic management of Barrett's esophagus: European Society of Gastrointestinal Endoscopy (ESGE) position statement. Endoscopy **49**(02), 191–198 (2017)
20. Zhang, L., et al.: Disentangling human error from ground truth in segmentation of medical images. In: Larochelle, H., Ranzato, M., Hadsell, R., Balcan, M., Lin, H. (eds.) Advances in Neural Information Processing Systems, vol. 33, pp. 15750–15762. Curran Associates, Inc. (2020). https://proceedings.neurips.cc/paper_files/paper/2020/file/b5d17ed2b502da15aa727af0d51508d6-Paper.pdf

Automated Hepatocellular Carcinoma Analysis in Multi-phase CT with Deep Learning

Krzysztof Kotowski[1], Bartosz Machura[1], Damian Kucharski[1,2],
Benjamín Gutiérrez-Becker[3], Agata Krason[4], Jean Tessier[4],
and Jakub Nalepa[1,2(✉)]

[1] Graylight Imaging, Gliwice, Poland
[2] Silesian University of Technology, Gliwice, Poland
Jakub.Nalepa@polsl.pl
[3] Roche, Pharma Research and Early Development,
Data and Analytics, Basel, Switzerland
[4] Roche Pharma Research and Early Development, Early Clinical Development Oncology, Roche Innovation Center Basel, Basel, Switzerland

Abstract. Hepatocellular carcinoma (HCC) is a common type of liver cancer. Its effective diagnosis and monitoring require analyzing computed tomography (CT) scans with intravenous contrast in multiple phases, taken at different intervals post-injection. Organ movement during these intervals, caused by factors like breathing, heartbeat, or patient motion, can affect the accuracy of HCC detection. Aligning two or more scans precisely, especially ensuring the liver's alignment, is crucial for reconstructing small lesions effectively. Additionally, the presence of various liver lesions, such as active HCC tumors, chemoembolizations, necrosis, portal vein thrombosis, cysts, or other lesions, complicates the diagnosis process. In this paper, we tackle these challenges and propose a deep learning pipeline for detecting, segmenting and ultimately quantifying HCC in multi-phase CT scans. Our rigorous experiments, conducted on a carefully curated dataset from a clinical trial involving HCC patients, demonstrate that our approach not only achieves high-quality detection and segmentation of HCC but also enables fully-automatic, objective, reproducible and accurate response assessment in HCC patients.

Keywords: Hepatocellular carcinoma · segmentation · response evaluation · RECIST · mRECIST · multi-phase CT

1 Introduction

Hepatocellular carcinoma (HCC) stands as one of the prevalent forms of liver cancer [17]. Studies indicate a connection between the occurrence of liver diseases like hepatitis B or C or cirrhosis and an elevated likelihood of developing

Supplementary Information The online version contains supplementary material available at https://doi.org/10.1007/978-3-031-73376-5_9.

HCC [5]. Nevertheless, HCC can also emerge in an otherwise healthy liver. In clinical settings, ultrasound, computed tomography (CT), or magnetic resonance imaging (MRI) scans are commonly used to assess the presence or probability of HCC in a patient [31]. Additionally, artificial intelligence (AI) algorithms, particularly those based on machine learning (ML), exhibit some capability in identifying HCC or predicting its onset from medical scans [4], especially for well-defined cohorts of patients [8,12,25]. However, the challenge in diagnosing HCC, either manually or automatically, arises from various factors, including the existence of different HCC subtypes, each with distinct enhancement patterns.

To effectively diagnose and monitor HCC, a concurrent examination of CT scan images obtained through intravenous contrast in multi-phase CT scans is necessary. These CT scans in different phases are captured at varying intervals following contrast injection. During these intervals, internal organs may shift in position and shape due to factors like breathing, heartbeat, or patient movement. Accurate detection of HCC from multiple scans requires precise alignment of two or more scans [14,27], particularly ensuring the liver's alignment within the images, crucial for effectively reconstructing small lesions. Furthermore, the presence of diverse classes of liver lesions, such as active HCC tumors, chemoembolizations within HCC tumors, necrosis, portal vein thrombosis, cysts, or other lesions, adds complexity to automatic or manual diagnosis. Therefore, there is a demand for enhanced systems for processing CT scan images to detect and predict HCC. We tackle these challenges—our contributions lay in proposing the first end-to-end pipeline for segmenting HCC in multi-phase CT scans, and for the automatic response assessment (Sect. 2), offering full reproducibility and objectiveness in tracking the disease. Our approaches are rigorously validated over meticulously curated CTs acquired for diverse HCC patients within a large-scale clinical study (Sect. 3). We believe that our efforts will become an important step toward ensuring unbiased assessment of response free from human error, leading to objective personalized clinical care. This is of paramount clinical importance in HCC analysis, as the response assessment involves detecting (extremely) small lesions which may be easily missed by a human.

2 Materials and Methods

2.1 Dataset

We build upon a dataset captured in a trial involving patients with HCC [23]. It included **184** patients, of which: **102** without cirrhosis, **82** with cirrhosis (33 females, 84 Asians, 78 Whites, 4 Afro-Americans, 18 unknown). The mean ($\pm$ std dev.) age was 66 ± 10 years (minimum: 34, maximum: 87). All CTs were acquired with ≤ 5 mm slice thickness, ≤ 5 mm reconstruction interval (distance between slice locations with contiguous slices). Three scans were acquired:

- **Scan 0**—pre-contrast scan (entire liver; this scan was optional),
- **Scan 1**—arterial phase scan (20–25 s post injection, entire liver),
- **Scan 2**—portal venous phase (60–70 s post injection, lung apices to symphysis pubis), and the delayed phase (min. 120 s post injection, entire liver).

2.2 Ground Truth Annotation Procedure

The dataset was annotated by 5 junior radiology specialists (3 years of experience on average) and 2 experts (20 years of experience on average). The experts reviewed the annotations prepared by the former. Each training set (T) annotation required acceptance from a single expert, whereas each test set (Ψ) annotation was double-checked and had to be agreed by both experts. The raters received coregistered CT scans, one arterial and one venous per patient, and contoured **five classes of objects (ground-truth lesion masks)**: (1) *active HCC tumors*, (2) *necrosis*, (3) *portal vein thrombosis*, (4) *cysts*, (5) *other lesions* (i.e., lesions which cannot be classified to the 1–4 classes). The raters specified the **confidence levels (1–4)** for each class (along with one confidence value of the overall segmentation) for every annotation. The confidence ranges from 1 to 4, and corresponds to *"I am not confident at all"*, *"I am not confident and require advice from a senior reader"*, *"I am confident to some extent, but I would prefer to have it reviewed by a senior reader"*, and *"I am fully confident"*, as suggested in [20]. The average value for all confidence values was higher than 3.5. The raters contoured chemoembolizations (with HU > 150, where HU denotes the Hounsfield units) too. Although the HCC is the most clinically important class, as these areas are used to identify and track the disease, we hypothesize that including other classes (*i*) can help improve the quality of HCC detection/segmentation, and (*ii*) may allow us to understand if HCC can be confused with e.g., cysts.

Table 1. Training-test dataset stratification.

Parameter→	V_t		N_t		V_{pv}		V_c	
Subset↓	Small	Big	Small	Big	Small	Big	Small	Big
Fold 1	14	18	18	14	26	6	24	8
Fold 2	15	17	18	14	26	6	24	8
Fold 3	17	15	18	14	26	6	25	7
Fold 4	17	15	18	14	26	6	25	7
Fold 5	17	15	16	16	26	6	25	7
Test (Ψ)	11	9	10	10	18	2	6	14

2.3 Training-Test Stratification Procedure

Four patients were removed from the dataset due to insufficient quality of coregistration according to the raters. The scans from the remaining **180 patients** were split into T and Ψ, with the former further divided into five non-overlapping folds. To ensure a similar distribution of tumor characteristics between subsets, patients were stratified according to: total volume of whole tumors (active HCC, chemoembolizations and necrosis combined) relative to the volume of the liver (V_t), total number of whole tumors (N_t), total volume of portal vein thrombosis (V_{pv}), and total volume of cysts (V_c). For every of these factors, each patient was assigned into one of two categories (*small* and *big*) based on the median value calculated for all lesions across all patients. Due to inconsistent characteristics of voxels and lack of representative samples, the *other lesions* class was not used in the stratification. Table 1 summarizes all subsets extracted from the entire dataset—we follow the patient-level splitting strategy to avoid training-test information leaks [24]. In this multi-fold strategy, each fold acts as a validation set (V) exactly once, e.g., during training of a base nnU-Net model.

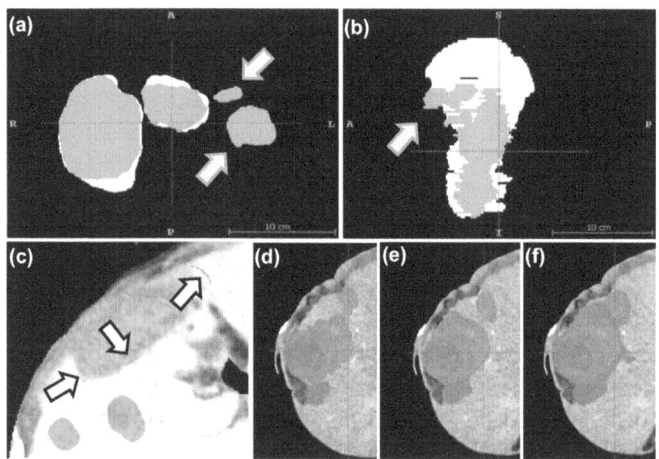

Fig. 1. Examples of (a–b) HCC ground-truth masks (in red) positioned outside the liver (in white) and (c) small ground-truth areas incorrectly annotated by the readers. In the ensemble segmentation model, setting the $|\mathcal{V}(Color figure online)|$ hyperparameter allows for adjusting the system's sensitivity: (d) ground-truth HCC, (e) $|\mathcal{V}| = 3$, (f) $|\mathcal{V}| = 1$.

2.4 Observed Issues and Further Preprocessing

The following issues concerning the ground-truth (GT) lesion masks were noted:

- Small parts of lesions were outside the liver, Fig. 1(a, b), e.g., for patients with advanced HCC or cirrhosis—liver masks and HCC GT were summed.
- HCC masks contained many extremely small tumors which were annotation mistakes or an effect of semi-automatic thresholding used by some raters, Fig. 1(c). All tumors smaller than $20\,\text{mm}^3$ were removed from the HCC GT. This threshold was chosen empirically by analyzing the volume distributions and confirmed by visual inspection of a senior radiologist.

The training set was preprocessed by cropping and masking the liver and by windowing voxels in the range of (-100 HU, 150 HU) [18]. The images were scaled to the (-1, 1.5) range, and the liver voxels were finally z-scored.

2.5 Automated HCC Analysis in Multi-phase CT

We introduce an end-to-end processing pipeline for automatic segmentation of liver tumors along with response assessment measurements. It contains three pivotal steps of coregistering arterial and venous CTs, segmenting HCC areas, and extracting unidimensional and volumetric characteristics of the lesions (Fig. 2). We employed distinct unidimensional response assessment algorithms based on the model output type to ensure robust evaluation of tumor progression. This included using RECIST (Response Evaluation Criteria in Solid Tumors 1.1 [28]; we use the fast AutoRANO [19]) for whole tumor (WT) comprising merged active HCC, chemoembolizations and necrosis, and mRECIST (Modified Response Evaluation Criteria in Solid Tumors [15,16]) for models predicting HCC.

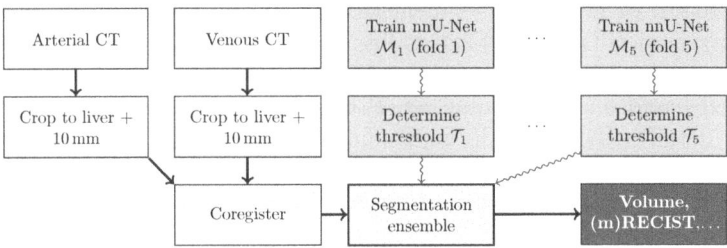

Fig. 2. Flowchart of the automated HCC analysis in multi-phase CT. The **thick black arrows** show the inference path, whereas the curly blue ones present the training path.

Coregistration of Arterial and Venous Phase CT Scans. The preprocessing steps were devised to standardize the region of interest (ROI) and eliminate extraneous data from arterial and venous CT scans prior to their coregistration. The initial step involved reorienting all scans (and segmentations) to the right, anterior, and superior (RAS) anatomical orientation [21]. Subsequently, the scans were cropped using AI-generated liver segmentation masks obtained using widely-established nnU-Net ("no-new-Net") deep learning models [9,10] (trained over the Liver Tumor Segmentation Benchmark (LiTS) [2]) to focus on the liver ROI, with an additional margin of 10 mm on each side. This margin is retained to preserve a small set of potential characteristic points in the vicinity of the liver, which could prove beneficial for the coregistration process[1]. To crop the scans to the liver, we:

1. **Dilate the liver segmentation mask** using an elliptical structuring element. The homogeneous margin initially defined in millimeters needs to be translated to voxel space. An elliptical kernel is used to accurately reflect the anisotropic spacings of the liver mask.

---

[1] In our initial study, we hypothesized that cropping scans around the liver could be advantageous by preserving essential anatomical structures and potential characteristic points like ribs, spine, or lungs, which are valuable for coregistration. However, coregistration sometimes overly prioritizes aligning anatomical structures unrelated to HCC—it may align bowels accurately while inadequately aligning the liver itself.

2. **Crop the scans to the dilated liver mask** without any additional margin.
3. **Mask all voxels outside the dilated liver segmentation mask** using a value of -1000, which corresponds to the air in Hounsfield units.

When designing coregistration, we must consider the natural range of organ deformations [7,22,26]. For example, individual bones are rigid, so rigid transformations suffice. The liver, being non-rigid, requires deformable coregistration with constraints due to its limited deformability. To ensure reproducibility, we exploit deformable coregistration and implement it using Greedy [30], with the normalized mutual information coregistration quality metric, 1.732×0.707 voxel coregistration regularization, and with the 0.5 warping step.

Segmentation of HCC. To segment HCC, we use the nnU-Net framework [9] to train five models ($\mathcal{M}_1$–$\mathcal{M}_5$) across five folds (Sect. 2.3). The architecture and parameters, including batch size, patch size, and voxel size, were automatically determined based on the training set T using the nnU-Net framework, as discussed in detail in [9], with the maximum number of epochs of 500. Its input comprises a CT with two coregistered, cropped, and preprocessed phases of identical shapes and voxel sizes. The output is a probability map that assigns tumor probabilities to each voxel, and the binarization thresholds were selected for each model separately. Here, we prioritized maximizing the Jaccard's index detection metric (JI), to ensure that small lesions are located. After identifying the threshold with the highest JI for V, we retain thresholds not significantly worse based on pairwise Wilcoxon tests ($p < 0.05$). From these, we select the threshold with the highest DICE. After selecting thresholds for each base model, we combined their predictions using a voting procedure. A voxel is classified as tumorous if it gets at least $|\mathcal{V}|$ votes ($1 \leq |\mathcal{V}| \leq 5$), with the $|\mathcal{V}|$ hyperparameter allowing to adjust the model's sensitivity, as shown in Fig. 1(d–f).

Training with Quality-Driven Weighted Loss. The average of DICE and categorical crossentropy is used as a loss function to train the base models, and apply the weighting procedure by using the tumor annotation quality confidence delivered by the senior radiologist (with the values of 1–4). Effectively, we weight the loss using the values of 0.75, 1.0, 1.25 or 1.5, respectively.

Quality Metrics. We use the *segmentation* and *detection* metrics. The former include DICE and Intersection-over-Union (IoU) calculated for the active part of the tumor not including chemoembolizations (i.e., not including voxels above 150 HU in both arterial and venous phase, as they are "trivial" to segment and would otherwise bias the metrics [3,29]). Although both metrics are used to evaluate small and big tumors, small ones have a marginal impact on these metrics. Detection metrics (precision, recall and the Jaccard's index) reflect how many lesions are localized correctly regardless of their size (a lesion is correctly detected if at least *one* voxel in prediction and GT overlaps)—they assess the performance of models in terms of tumor identifications. All detection metrics are calculated for the whole tumor. All metrics range from 0 to 1, with one indicating the perfect score.

3 Experimental Study

The objectives of the study were two-fold: (*i*) to investigate specialized models analyzing the WT (active HCC, chemoembolizations, and NEC, where NEC denotes necrosis), and (*ii*) to verify the quality of automatic response measurements. The algorithm-rater agreement was evaluated using the Intraclass Correlation Coefficient (ICC) calculated on a single measurement, absolute-agreement, two-way random-effects model. The R package IRR (Inter Rater Reliability, v. 0.84.1) was used for ICC, and GraphPad Prism 10.2.0 was used for statistical analysis.

Table 2. Segmentation and detection metrics obtained using the investigated models. The best metrics for the models targeting HCC detection and segmentation are **bold**.

Model	Class	Segmentation					Detection		
		DICE (↑)	Sen (↑)	Spe (↑)	H95 (↓)	IoU (↑)	JI (↑)	Pre (↑)	Re (↑)
WT	WT	0.774	0.754	0.997	24.471	0.654	0.382	0.506	0.506
HCC	HCC	**0.735**	**0.721**	**0.997**	19.242	**0.601**	**0.378**	**0.507**	**0.466**
HCC+NEC	HCC	**0.735**	0.707	**0.997**	**18.034**	0.598	0.306	0.462	0.367
HCC+NEC	NEC	0.403	0.420	0.999	22.853	0.289	0.169	0.228	0.254

3.1 Results and Discussion

Table 2 gathers the results obtained over the unseen test set Ψ using all investigated models, trained over GT containing WT, HCC, as well as the HCC and NEC areas (HCC+NEC; see the *Model* column). Response assessment measurements are commonly calculated for either WT or HCC regions, and the automated segmentation of these regions is significantly better than for NEC—this is also reflected in the lowest Hausdorff distance values (H95) obtained for HCC. Interestingly, including the NEC class in training did not improve the quality of the models targeting HCC. Figure 3 displays Bland-Altman plots, indicating strong agreement between HCC and HCC+NEC models and GT with only few outliers (large disagreements between GT and predictions). The model consistently underestimates predicted lesions compared to GT, as evidenced by most points clustering on one side of the zero difference axis in Bland-Altman plots. Also, volumetric measurements calculated for prediction masks exhibit strong correlation with GT volume, offering robustness in tracing cancerous changes in time (Table 3). RECIST GT exhibits the lowest Pearson's correlation coefficients, suggesting it may not reliably track lesion or assess tumor burden. In contrast, all values on prediction masks show strong correlation with the volume calculated for GT. With the highest Pearson's coefficients for volumetric measurements reaching 0.989 (prediction vs. GT), the model demonstrates accuracy in reflecting cancerous changes in multi-phase CT. Of note, the results showed that HCC was virtually never confused with other liver lesions.

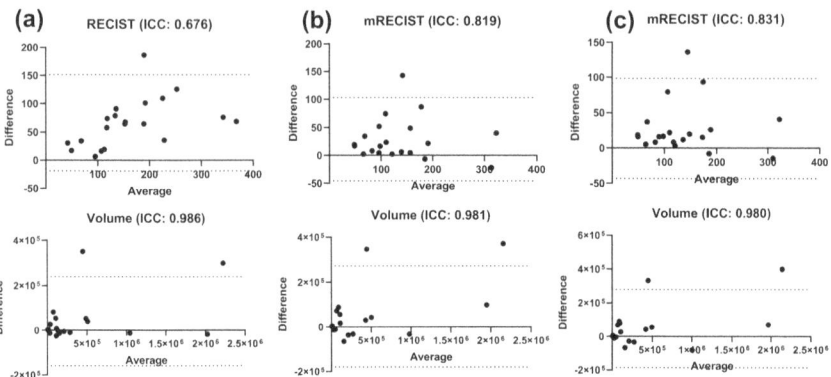

Fig. 3. Bland-Altman plots of agreement between RECIST/mRECIST and volume obtained using the (a) WT, (b) HCC, and (c) HCC+NEC models in relation to GT.

Table 3. Pearson's correlation coefficient (↑) across different metrics and models.

Model→	HCC			HCC+NEC		
Metric	V_{GT}	RECIST	V_{H+N}	V_{GT}	RECIST	V_{H+N}
$RECIST_{GT}$	0.787	0.882	0.766	0.787	0.890	0.767
V_{GT}		0.904	0.989		0.909	0.987
$RECIST_{H+N}$			0.906			0.900

3.2 Limitations

We acknowledge several limitations in our study. While we rigorously validated our techniques on a carefully curated Ψ from a large-scale clinical trial, additional verification on diverse datasets would strengthen the findings. We wanted to do this in our study, but there are no public sets including both arterial and venous CT phases—even the well-established LiTS [2] and Medical Decathlon [1] have only venous phases, so it is impossible to assess our approach over them. Although we assessed the quality of all steps of our pipeline, implementing additional quality control [6] and explanation techniques [11,13] could enhance the utility of the system. They could highlight the most significant (or suspicious) parts of the scans, thereby allowing clinicians to "look into the black AI box" or halt the analysis for e.g., incorrectly coregistered or corrupted scans.

3.3 Clinical Relevance and Impact

Our pipeline for comprehensive analysis of HCC in multi-phase CT scans achieves superior detection, segmentation, and quantifiable characteristic extraction of HCC patients. It ensures fully-automatic, reproducible, and accurate response assessment in HCC patients which may play a key role in clinical settings to objectively detect and track the disease progression without any human bias.

4 Conclusion

Hepatocellular carcinoma, a prevalent type of liver cancer, demands precise analysis and monitoring for effective clinical management. Our study presents the first end-to-end, fully-automated and reproducible pipeline for HCC segmentation and response assessment from multi-phase CT scans. By showcasing its effectiveness and robustness through an array of thorough experiments performed on a meticulously curated dataset from a large-scale clinical trial and backed up with rigorous statistical testing, we believe that our work represents a substantial step forward in achieving objective personalized cancer care.

Acknowledgments. JN was supported by the Silesian University of Technology funds through the grant for maintaining and developing research potential. The authors would like to thank Marek Pitura and Karolina Adamowska (Graylight Imaging) for their valuable help in managing this study.
This paper is in memory of Dr. Grzegorz Nalepa, an extraordinary scientist, pediatric hematologist/oncologist, and a compassionate champion for kids at Riley Hospital for Children, Indianapolis, USA, who helped countless patients and their families through some of the most challenging moments of their lives.

Disclosure of Interests. The authors declare the following financial interests/personal relationships which may be considered as potential competing interests: Benjamín Gutiérrez-Becker has ownership in Roche and Jean Tessier is Head of Clinical Imaging at Roche pRED Oncology and has ownership in Roche.

References

1. Antonelli, M., Reinke, A., Bakas, S., et al.: The medical segmentation decathlon. Nat. Commun. **13**(1), 4128 (2022). https://doi.org/10.1038/s41467-022-30695-9
2. Bilic, P., Christ, P., Li, H.B., et al.: The Liver Tumor Segmentation benchmark (LiTS). Med. Image Anal. **84**, 102680 (2023). https://doi.org/10.1016/j.media.2022.102680
3. Bryant, M.K., Dorn, D.P., Zarzour, J., et al.: Computed tomography predictors of hepatocellular carcinoma tumour necrosis after chemoembolization. HPB **16**(4), 327–335 (2014). https://doi.org/10.1111/hpb.12149
4. Calderaro, J., Seraphin, T.P., Luedde, T., Simon, T.G.: Artificial intelligence for the prevention and clinical management of hepatocellular carcinoma. J. Hepatol. **76**(6), 1348–1361 (2022). https://doi.org/10.1016/j.jhep.2022.01.014
5. El-Serag, H.B.: Epidemiology of viral hepatitis and hepatocellular carcinoma. Gastroenterology **142**(6), 1264-1273.e1 (2012). https://doi.org/10.1053/j.gastro.2011.12.061
6. Fournel, J., Bartoli, A., Bendahan, D., et al.: Medical image segmentation automatic quality control: a multi-dimensional approach. Med. Image Anal. **74**, 102213 (2021). https://doi.org/10.1016/j.media.2021.102213
7. Fukumitsu, N., Nitta, K., Terunuma, T., et al.: Registration error of the liver CT using deformable image registration of MIM Maestro and Velocity AI. BMC Med. Imaging **17**(1), 30 (2017). https://doi.org/10.1186/s12880-017-0202-z

8. Ioannou, G.N., Tang, W., Beste, L.A., et al.: Assessment of a deep learning model to predict hepatocellular carcinoma in patients with hepatitis C cirrhosis. JAMA Netw. Open **3**(9), e2015626–e2015626 (2020). https://doi.org/10.1001/jamanetworkopen.2020.15626
9. Isensee, F., Jaeger, P.F., Kohl, S.A.A., et al.: nnU-Net: a self-configuring method for deep learning-based biomedical image segmentation. Nat. Methods **18**(2), 203–211 (2021). https://doi.org/10.1038/s41592-020-01008-z
10. Isensee, F., Kickingereder, P., Wick, W., et al.: No new-net. In: Crimi, A., Bakas, S., Kuijf, H., Keyvan, F., Reyes, M., van Walsum, T. (eds.) Brainlesion: Glioma, Multiple Sclerosis, Stroke and Traumatic Brain Injuries, pp. 234–244. Springer International Publishing, Cham (2019). https://doi.org/10.1007/978-3-030-11726-9_21
11. Jin, Z., Pang, M., Yang, Y., et al.: Explaining massive-training artificial neural networks in medical image analysis task through visualizing functions within the models. In: Greenspan, H., et al. (eds.) Medical Image Computing and Computer Assisted Intervention – MICCAI 2023, pp. 713–722. Springer, Cham (2023). https://doi.org/10.1007/978-3-031-43895-0_67
12. Kim, H.Y., Lampertico, P., Nam, J.Y., et al.: An artificial intelligence model to predict hepatocellular carcinoma risk in Korean and Caucasian patients with chronic hepatitis B. J. Hepatol. **76**(2), 311–318 (2022). https://doi.org/10.1016/j.jhep.2021.09.025
13. Komorowski, P., Baniecki, H., Biecek, P.: Towards evaluating explanations of vision transformers for medical imaging. In: Proceedings of the IEEE/CVF Conference on Computer Vision and Pattern Recognition (CVPR) Workshops, pp. 3726–3732 (2023)
14. Kulkarni, N.M., Fung, A., Kambadakone, A.R., Yeh, B.M.: Computed tomography techniques, protocols, advancements, and future directions in liver diseases. Magn. Reson. Imaging Clin. N. Am. **29**(3), 305–320 (2021). https://doi.org/10.1016/j.mric.2021.05.002
15. Lee, J.S., Choi, H.J., Kim, B.K., et al.: The Modified Response Evaluation Criteria in Solid Tumors (RECIST) yield a more accurate prognoses than the RECIST 1.1 in hepatocellular carcinoma treated with transarterial radioembolization. Gut Liver **14**(6), 765–774 (2020). https://doi.org/10.5009/gnl19197
16. Lencioni, R., Llovet, J.M.: Modified RECIST (mRECIST) assessment for hepatocellular carcinoma. Semin. Liver Dis. **30**(01), 052–060 (2010). https://doi.org/10.1055/s-0030-1247132
17. Llovet, J.M., Kelley, R.K., Villanueva, A., et al.: Hepatocellular carcinoma. Nat. Rev. Dis. Primers. **7**(1), 6 (2021). https://doi.org/10.1038/s41572-020-00240-3
18. Mayo-Smith, W.W., Gupta, H., Ridlen, M.S., et al.: Detecting hepatic lesions: the added utility of CT liver window settings. Radiology **210**(3), 601–604 (1999). https://doi.org/10.1148/radiology.210.3.r99mr07601, pMID: 10207455
19. Nalepa, J., Kotowski, K., Machura, B., et al.: Deep learning automates bidimensional and volumetric tumor burden measurement from MRI in pre- and postoperative glioblastoma patients. Comp. Biol. Med. **154**, 106603 (2023)
20. Nalepa, J., Ribalta Lorenzo, P., Marcinkiewicz, M., et al.: Fully-automated deep learning-powered system for DCE-MRI analysis of brain tumors. Artif. Intell. Med. **102**, 101769 (2020)
21. Rood, J.E., et al.: Toward a common coordinate framework for the human body. Cell **179**(7), 1455–1467 (2019). https://doi.org/10.1016/j.cell.2019.11.019, https://www.sciencedirect.com/science/article/pii/S0092867419312759

22. Sarudis, S., Karlsson, A., Bibac, D., et al.: Evaluation of deformable image registration accuracy for CT images of the thorax region. Physica Medica **57**, 191–199 (2019). https://doi.org/10.1016/j.ejmp.2018.12.030
23. Finn, R.S., et al.: Atezolizumab plus bevacizumab in unresectable hepatocellular carcinoma. N. Engl. J. Med. **382**(20), 1894–1905 (2020). https://doi.org/10.1056/NEJMoa1915745
24. Wijata, A.M., Nalepa, J.: Unbiased validation of the algorithms for automatic needle localization in ultrasound-guided breast biopsies. In: 2022 IEEE International Conference on Image Processing (ICIP), pp. 3571–3575 (2022)
25. Wong, G.L.H., Hui, V.W.K., Tan, Q., et al.: Novel machine learning models outperform risk scores in predicting hepatocellular carcinoma in patients with chronic viral hepatitis. JHEP Rep. **4**(3), 100441 (2022)
26. Woolcot, T., Kousi, E., Wells, E., et al.: An evaluation of systematic errors on marker-based registration of computed tomography and magnetic resonance images of the liver. Phys. Imaging Radiat. Oncol. **7**, 27–31 (2018)
27. Wu, L., Wang, H., Chen, Y., et al.: Beyond radiologist-level liver lesion detection on multi-phase contrast-enhanced CT images by deep learning. iScience **26**(11), 108183 (2023). https://doi.org/10.1016/j.isci.2023.108183
28. Yu, H., Bai, Y., Xie, X., et al.: RECIST 1.1 versus mRECIST for assessment of tumour response to molecular targeted therapies and disease outcomes in patients with hepatocellular carcinoma: a systematic review and meta-analysis. BMJ Open **12**(6) (2022). https://doi.org/10.1136/bmjopen-2021-052294
29. Yuan, Z., Ye, X.D., Dong, S., et al.: Evaluation of early imaging response after chemoembolization of hepatocellular carcinoma by phosphorus-31 magnetic resonance spectroscopy-initial experience. J. Vasc. Interv. Radiol. **22**(8), 1166–1173 (2011)
30. Yushkevich, P.A., Pluta, J., Wang, H., et al.: Fast automatic segmentation of hippocampal subfields and medial temporal lobe subregions in 3 tesla and 7 tesla t2-weighted MRI. Alzheimer's Dementia **12**(7S_Part_2), P126–P127 (2016)
31. Zerunian, M., Di Stefano, F., Bracci, B., et al.: Imaging of Hepatocellular Carcinoma. In: Ettorre, G.M. (eds.), pp. 37–43. Springer International Publishing, Cham (2023). https://doi.org/10.1007/978-3-031-09371-5_5

Refining Deep Learning Segmentation Maps with a Local Thresholding Approach: Application to Liver Surface Nodularity Quantification in CT

Sisi Yang[1,2,3(✉)], Alexandre Bône[1], Thomas Decaens[4], and Joan Alexis Glaunès[2]

[1] Guerbet Research, Villepinte, France
sisi.yang@guerbet.com
[2] MAP5 (UMR 8145), Université Paris Cité, Paris, France
[3] Radiologie B, Hôpital Cochin, Paris, France
[4] Centre Hospitalier Universitaire Grenoble-Alpes, Paris, France

Abstract. Liver fibrosis is a chronic disease that must be treated to prevent further complications, including liver cancer. The diagnosis of liver fibrosis in CT imaging can be challenging and is often subject to disagreements between radiologists. The nodularity of the liver surface is a well-known feature of fibrosis, which can be quantified in clinical practice with specialized software applications that rely on semi-automatic delineation of the liver contours. This approach, however, requires a high degree of expertise and is time-consuming. While deep learning methods have recently shown excellent performance for liver segmentation, the predicted contours are typically insufficiently accurate for nodularity quantification. In this work, we propose a local thresholding approach to refine the predictions of a deep network trained to segment the liver in CT images. We show that our refinement method improves the estimation of the liver surface nodularity compared to a baseline deep network, with Spearman's correlation coefficients of 0.60 and 0.47, respectively. This new estimator predicts advanced fibrosis better than the reference clinical approach, with areas under the curve of 74.6% and 67.9%, respectively.

Keywords: Liver fibrosis · surface nodularity · biomarkers · segmentation · post-processing · explainable artificial intelligence

1 Introduction

Liver fibrosis is a chronic liver disease, that can evolve into cirrhosis and favor the development of primary liver cancer if left untreated. Liver cancer is a leading

---

Supplementary Information The online version contains supplementary material available at https://doi.org/10.1007/978-3-031-73376-5_10.

cause of mortality worldwide, and a recognized global public health issue [15]. Biopsy remains the gold standard for diagnosing and staging liver fibrosis but is invasive and carries risks [6]. Noninvasive methods such as serum tests (e.g., Fibrotest, Fibrometers) and liver stiffness measurement (e.g., transient elastography) are available but have limitations in accuracy [8] and applicability [3].

Healthy livers present a smooth surface, whereas fibrotic livers have an irregular surface. This irregularity is a well-known imaging feature of fibrosis, and can be visually assessed using ultrasound, CT or MR images [1]. In [17], a quantitative 'liver surface nodularity' (LSN) score was proposed, which strongly correlated with the fibrosis stage. Radiologists measure this fibrosis biomarker using interactive software tools [5], which require users to manually define regions of interest in well-chosen segments of the liver surface. A thresholding approach then automates liver contour delineation, but this method is time-consuming, requiring radiologists to balance speed and accuracy [16].

In recent years, deep learning approaches became state-of-the-art for many biomedical segmentation tasks [7]. Building on the seminal UNet model [14], the nnUNet framework defined a series of preprocessing, architecture, training and inference procedure heuristics that proved sufficient to top a variety of biomedical image segmentation benchmarks [9], including liver-specific challenges like LITS [2]. Deep segmentation networks typically use overlap-based loss functions like the Dice coefficient [11], which value the contribution of each segmented voxel equally, regardless of its topological nature, such as whether it is on the object's boundary or not. Additionally, reference segmentation maps are typically manually drawn using 'paint brush'-like interactive tools available in specialized software like ITK-SNAP [2,20], which inherently involve the approximation of object contours. Consequently, while deep learning has effectively addressed automatic volume estimation, it lacks accuracy in surface estimation and liver surface nodularity quantification. We further argue that in favorable areas of the image, classical, learning-free segmentation approaches like histogram thresholding [12,13] or active contour methods [4] may outperform deep learning for liver surface delineation and nodularity quantification.

In this study, we propose a post-processing method that refines deep learning segmentation maps using a local thresholding approach. This method consists of a set of heuristics that allow us to identify the well-contrasted regions of the image, where learning-free segmentation can be advantageously leveraged. The method was evaluated on a test database of abdominal CT scans. First, 'liver surface nodularity' (LSN) scores were computed from initial and refined liver contours, and compared to the reference LSN scores obtained with a semi-automatic clinical software application. Second, the predictive power of these LSN scores was assessed for liver fibrosis grading. Finally, the sensitivity of the proposed method to hyperparameters was studied in a final experiment.

2 Liver Contour Refinement Method

The proposed liver contour refinement method relies on the hypothesis that, in well-chosen regions of the scan images, simple thresholding, i.e., learning-free segmentation, may be more accurate than deep learning for liver surface delineation. This fundamental claim is illustrated by the top row of Fig. 1, which shows examples of liver contour results from the methods of [9] ('nnUNet') and [17] – as implemented in [5] (Liver Boundary Analysis v0.88, 'LBA'), two state-of-the-art methods in their respective communities. We can see that while the nnUNet method automatically and robustly computes a convincing liver segmentation when considered as a whole, the local surface delineation is less accurate than the contour semi-automatically obtained with the LBA software. The LBA method requires an expert user to manually define regions of interest with a 'paint brush' interactive tool before local thresholding can be applied.

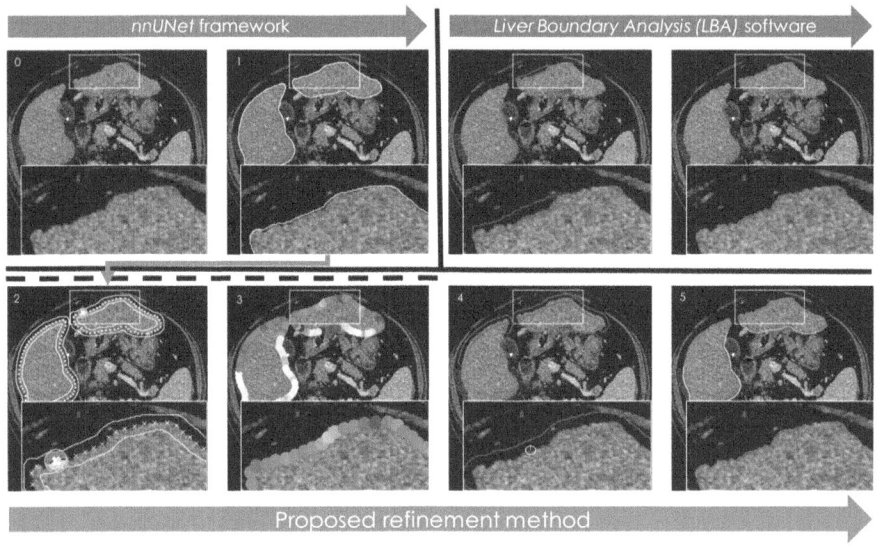

Fig. 1. Illustration of two state-of-the-art methods for liver segmentation in CT (top row), and of the proposed liver contour refinement method (bottom row) with CT images of a randomly selected cirrhotic patient.

The bottom row of Fig. 1 illustrates the proposed liver contour refinement pipeline, which aims to combine the best of both aforementioned approaches. The following steps are executed successively, illustrated by numbered images in Fig. 1. The images without numbers are reproductions of the LBA software.

1. **Initial segmentation.** Segment the entire liver in an abdominal CT image using a pretrained deep network, e.g., [9]. This method is expected to be accurate for volume estimation, but approximate for surface delineation (image n° 1) .
2. **Liver shell.** Using dilation and erosion operators, e.g., from scikit-image [19], compute thick outer and inner 'liver shell' binary segmentation maps. This

step involves a radius hyperparameter R_{shell}, expressed in millimeters, represented by red dotted lines (image n° 2)

3. **Contrast map.** At each point along the liver contour, intersect a disk with the outer and inner liver shells to define two corresponding regions. Compute the local contrast at that point, defined as the difference in median intensity between these two regions inside the disk. Collect local contrast values to build the contrast map (image n° 3). This step involves a radius hyperparameter $R_{contrast}$ (millimeters), represented by the red dotted line inside of the disk within which the contrast is calculated. (image n° 2)
4. **High-contrast regions.** Threshold the contrast map to identify the high-contrast segments of the liver contour, i.e. where the delineation between liver parenchyma and body fat is sufficiently conspicuous. Using the distance transform, determine the neighboring region of interest (ROI) for each segment. This step involves a threshold hyperparameter Δ_{min} (Hounsfield units) and a radius hyperparameter R_{roi} (millimeters), the white circle of radius R_{roi} represents the disk within which the algorithm searches for contours around each point of the deep-learning contours (image n° 4).
5. **Segmentation refinement.** In each high-contrast region, segment the liver contour using a thresholding-based approach. We used Otsu's method [12] (image n° 5).

In the rest of this article, the hyperparameters were fixed to $R_{shell} = R_{contrast} = 5mm$, $R_{roi} = 2.5mm$, and $\Delta_{min} = 60HU$, unless specified otherwise. These hyperparameters encode fundamental anatomical and biophysical characteristics of the human liver, such as the typical geometry of fat distribution and x-ray absorption coefficients. Therefore, these hyperparameters could reasonably generalize to other CT acquisition protocols.

Implementation Details. The described pipeline from steps 2 to 5 was implemented in Python, using scikit-image implementation of Otsu's method for the final step. Code available at: https://github.com/Guerbet-AI/auto-LSN All geometrical operators were applied in 2D on a slice-per-slice basis. The liver contour refinement pipeline typically runs in about 30 s on the 4-core Intel i5-1145G7 processor to post-process thin-slice liver segmentation.

3 Experiments

3.1 Databases

For the baseline deep segmentation network, a private database of 3270 abdominal portal-phase CT scans from 1975 patients and the public LITS test database [2] were leveraged as train and test sets, respectively. Both databases were multicentric but without overlap, and the patients had a variety of liver diseases. Coarse liver segmentation maps, manually defined by experienced radiologists, were systematically available. For the proposed liver contour refinement method, a private monocentric database of 102 portal-phase images acquired on different CT scanner models (See supplementary material) from as many patients was leveraged as a test set. 88 were male, and the mean age (± the standard deviation) was 72 years (±10 years). All patients were suspected of primary hepatic tumors and underwent a liver biopsy with an available Metavir histological score. According to this score, 13 patients had no fibrosis (F0 stage), 7 had mild or moderate fibrosis (F1 or F2), 18 had severe fibrosis (F3), and 57 had cirrhosis (F4).

3.2 Quantification of the Liver Surface Nodularity (LSN Score)

Firstly, the nnUNet framework [9] was leveraged to train a baseline liver segmentation deep network on the private database detailed in Sect. 3.1. All data preprocessing, model architecture, training and inference hyperparameters were automatically determined following the nnUNet (v1) heuristic rules. The liver segmentation method achieved a 95.5% average Dice score on the LITS test set. To acquire reference LSN measurements, a radiologist (S.Y, 8 years of experience), blinded to all clinical data, reviewed all 102 test scans using the Liver Boundary Analysis software (LBA v0.88). Figure 2 shows a manual ROI drawing with an interactive 'paint brush' tool from LBA v0.88 (left panel). Following the user manual, ROIs were placed on the anterior left lobe or lateral right lobe, where the liver is clearly outlined by fat or effusion. The software then automatically delineated the liver contour and computed a case-level LSN score ('LBA-LSN') as described in [17] (right panel in Fig. 2). Following [16], a minimum of 8 measurements were obtained per case. Reference manual segmentations for liver contours were unavailable; only reference LSN measurements (one scalar value per test patient) were available for evaluation.

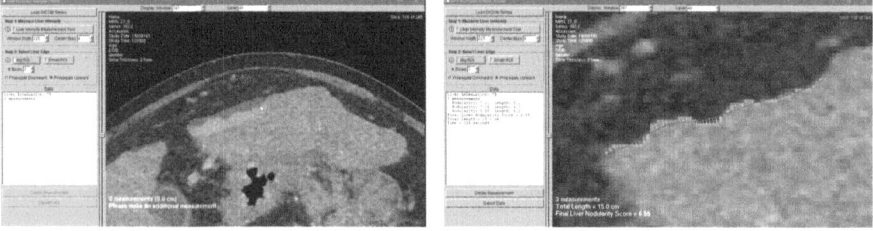

Fig. 2. Snapshots from the Liver Boundary Analysis software. First, a region of interest was manually defined (left panel). Second, the liver contours were automatically computed.

Following the same procedure, other LSN measurements were automatically computed from two sets of liver contours: (i) the baseline contours predicted by the nnUNet model ('nn-LSN') which correspond to step 1 in Sect. 2, (ii) the refined contours computed as proposed in Sect. 2 ('auto-LSN').

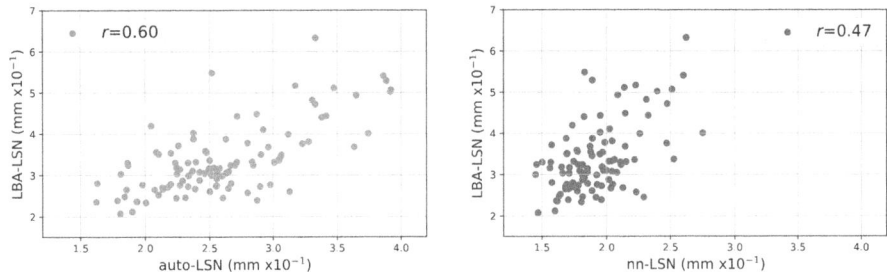

Fig. 3. Associations between the liver surface nodularity scores obtained with auto-LSN and nn-LSN, with respect to the reference LBA-LSN approach.

Performance Metrics and Statistical Analysis. The Spearman rank correlation coefficient (r) was used to assess the association with reference LSN measurements. Confidence intervals were computed via 10000-repetition bootstrapping. The intraclass correlation coefficient (ICC) was used to evaluate variability among reference LSN measurements (LBA-LSN).

Results. Scatter plots corresponding to these two configurations are shown in Fig. 3. The association scores were 0.47 (95% CI [0.29, 0.62]) for the baseline and 0.60 (95% CI [0.44, 0.74]) for the refined liver contours, with the difference being significant (p = 0.03), suggesting that our contour refinement method significantly improved liver surface delineation. The ICC of the LBA-LSN score was 0.78 using a single measurement as the basis, 0.98 using the mean of measurements as the basis.

3.3 Classification of Liver Fibrosis (Metavir Score)

Metavir scores were used as reference liver fibrosis labels. Two partitions of the test patients were defined, corresponding to two Metavir score binarizations. The first partition opposed the 'cirrhosis' group (Metavir stage F4, N = 57) to the non-cirrhosis group (N = 45). The second partition similarly opposed the 'advanced fibrosis' group (F3 or F4, N = 75) to the remaining cases (F0, F1 or F2, N = 27).

The predictive power of LSN measurements for fibrosis classification was assessed in two configurations (Sect. 3.2): (i) semi-automatically defined LSN measurements with the Liver Boundary Analysis software ('LBA-LSN'), and (ii) automatically computed LSN measurements from the refined nnUNet liver contours ('auto-LSN').

The volume of the liver left lateral segment normalized by the total liver volume ('LLS-TLV') is a well-known biomarker of liver fibrosis [10], and served as baseline biomarker in our study. Left lobe segmentation maps were automatically computed for the 102 test cases following the method of [18] and systematically reviewed by a radiologist (S.Y, 8 years of experience). The LLS-TLV ratios were then derived using the nnUNet predictions for normalization.

Performance Metrics and Statistical Analysis. Areas under the curve (AUCs) were assessed as performance metric, and DeLong's test was used for their comparisons. Their respective 95% CI were calculated.

Results. Figure 4 illustrates receiver operating characteristic curves for the three aforementioned biomarkers. AUCs were 69.4% (95% CI [59.0, 79.5]), 68.8% (95% CI [58.0, 78.8]), and 61.1% (95% CI [49.8, 71.9]) for LBA-LSN, auto-LSN, and LLS-TLV, respectively, in the cirrhosis classification. DeLong's test yielded p-values of 0.90 and 0.27 when comparing auto-LSN with LBA-LSN and LLS-TLV, respectively. Therefore, there was no significant difference of performance between these three biomarkers for the diagnosis of cirrhosis. For advanced fibrosis, AUCs were 67.9% (95% CI [55.5, 79.2]), 74.6% (95% CI [61.6, 86.3]), and 71.8% (95% CI [58.5, 83.8]) for LBA-LSN, auto-LSN, and LLS-TLV, with p-values of 0.25 and 0.72 when comparing auto-LSN with LBA-LSN and LLS-TLV, respectively. Although auto-LSN achieved a higher AUC compared to the reference LBA-LSN, this difference was not statistically significant.

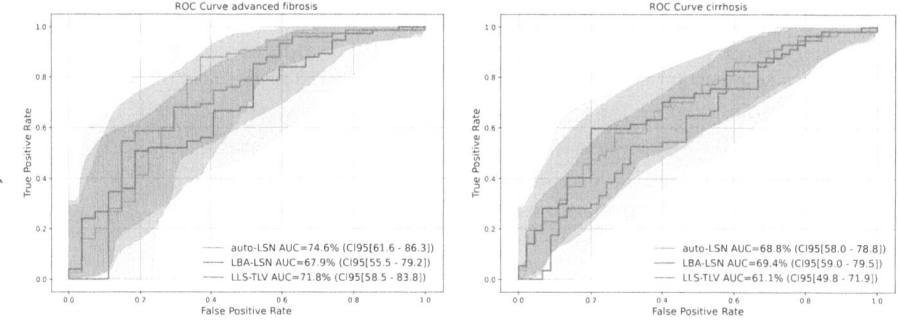

Fig. 4. Receiver operating characteristic curves for advanced fibrosis and cirrhosis classification. The proposed auto-LSN is compared to the clinical reference LBA-LSN and the baseline LLS-TLV.

3.4 Robustness to Hyperparameter Selection

The hyperparameters proposed in Sect. 2 were selected based on the authors' experience in reading hepatic CT scans. Their respective influence on LSN quantification and fibrosis classification performance was evaluated by successively perturbing their original values while keeping other hyperparameters constant. Figure 5 reports the resulting Spearman's r and AUC metrics.

4 Discussion and Conclusion

We proposed a method to refine predictions of a deep segmentation network for quantifying liver surface nodularity, a critical fibrosis biomarker. First, on a test

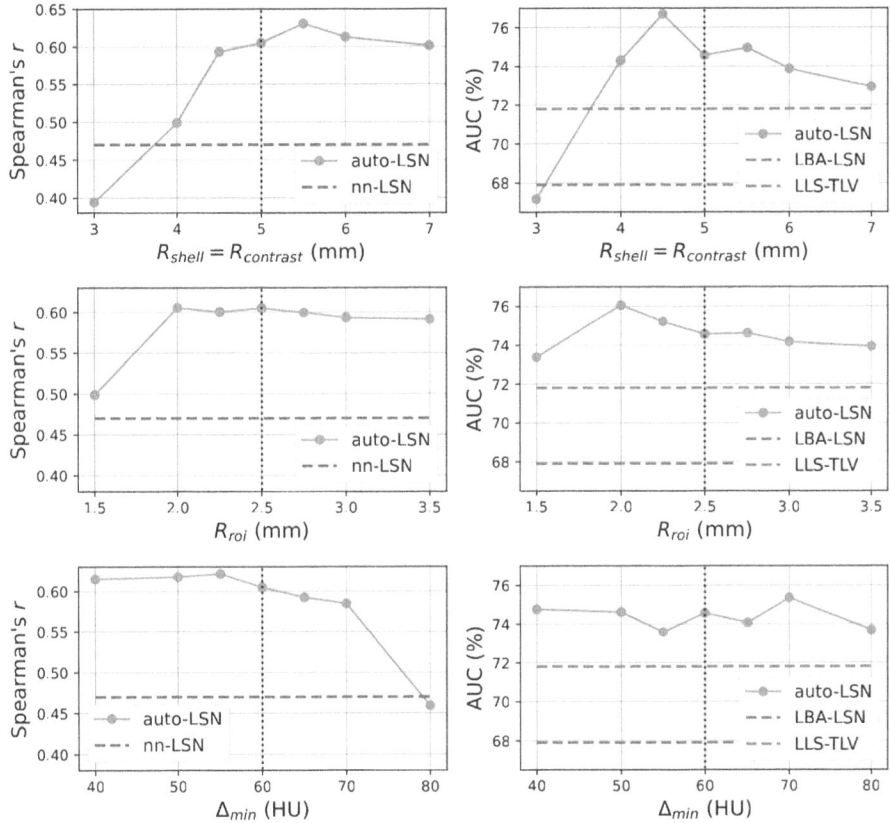

Fig. 5. Evaluation of the method's robustness to hyperparameter selection. Black vertical dotted lines mark the initial hyperparameter values from Sect. 2.

database of 102 cases, the refined segmentation maps proved better-suited for liver surface nodularity quantification than the initial predictions of a nnUNet model calibrated to segment the liver on a large training database of 3270 CT scans. A superior correlation was achieved (r = 0.60 versus r = 0.47, p=0.03), with respect to reference nodularity scores obtained by an experienced radiologist with a specialized clinical software application. Our refinement approach could not be more directly evaluated by comparing the resulting segmentation maps with reference manually-defined counterparts, as we argue that most common annotation protocols and tools necessarily lead to a non-negligible degree of approximation in contour delineation. The nnUNet could be improved by adding a strong, fully deep-learning-based method specialized in the contour delineation task, using a weight map to penalize the model if the contour delineation is not accurate, or by adding an L1 loss term. Such models were deliberately excluded from the present study, for two main reasons: (i) the lack of interpretability of the resulting models, which may restrict their acceptance for clinical use,(ii) the cost associated with train set building. Furthermore, nodularity scores derived from the refined segmentation maps exhibited enhanced predictive power for liver fibrosis grading compared to the reference semi-automatic approach, achieving AUCs of 74.6% and 67.9% respectively, for advanced fibrosis classification. This represents a notable clinical advancement by enabling fully automatic extraction of this liver fibrosis biomarker in routine CT imaging. Our method analyzes all slices covering the liver volume in 30 s per patient, whereas the reference semi-automatic approach requires more than one minute on average for only a few slices [16], therefore it is more time-efficient and more representative of the entire liver health. It requires no manual input, being more objective, reducing inter-reader variability in follow-up scenarios. This algorithm could be integrated into the medical image viewer to provide radiologists with a systematic, automated LSN score when interpreting abdominal scans in clinical routine. It could also be applied retrospectively to previous imaging studies of the patients.

Our algorithm has limitations, particularly in CT slices with poor liver contour contrast (See material). However, it generated auto-LSN scores consistently for all patients in our cohort without outliers, with the output being the median of the distances on all slices. Another limitation of our approach is the necessity to select hyperparameters. The influence of hyperparameters was evaluated in the last experiment. All the previously reported results held when the original hyperparameters were slightly modified. Nonetheless, excessive perturbations may lead to non-negligible loss of performance. Future research aims to extend our 2D refinement approach to 3D, potentially improving the extraction of predictive fibrosis biomarkers from liver surfaces.

Acknowledgements. This work was supported by the French National Research Agency (ANR) as part of the Investments for the Future programme (PIA) under grant agreement ANR-21-RHUS-01.

Disclosure of Interests. S. Yang and A. Bône are employed by Guerbet.

References

1. Berzigotti, A., et al.: Ultrasonographic evaluation of liver surface and transient elastography in clinically doubtful cirrhosis. J. Hepatol. **52**(6), 846–853 (2010). https://doi.org/10.1016/j.jhep.2009.12.031
2. Bilic, P., et al.: The liver tumor segmentation benchmark (LITS). Med. Image Anal. **84**, 102680 (2023). https://doi.org/10.1016/j.media.2022.102680
3. Castera, L.: Noninvasive methods to assess liver disease in patients with hepatitis B or C. Gastroenterology **142**(6), 1293–1302 (2012). https://doi.org/10.1053/j.gastro.2012.02.017
4. Chan, T., Vese, L.: An active contour model without edges. In: International Conference on Scale-space Theories in Computer Vision, pp. 141–151. Springer (1999). https://doi.org/10.1007/3-540-48236-9_13
5. Food, U., Administration, D.: 510(k) summary: K201092. https://www.accessdata.fda.gov/cdrh_docs/pdf20/K201092.pdf (2020). Accessed 19 Jun 2024

6. Glaser, J., Pausch, J.: The risk of liver biopsy. Z. Gastroenterol. **33**(11), 673–676 (1995)
7. Gul, S., Khan, M.S., Bibi, A., Khandakar, A., Ayari, M.A., Chowdhury, M.E.: Deep learning techniques for liver and liver tumor segmentation: A review. Comput. Bio. Med. **147**, 105620 (2022). https://doi.org/10.1016/j.compbiomed.2022.105620
8. Holmberg, S.D., et al.: Noninvasive serum fibrosis markers for screening and staging chronic hepatitis C virus patients in a large us cohort. Clin. Infect. Dis. **57**(2), 240–246 (2013). https://doi.org/10.1093/cid/cit245
9. Isensee, F., Jaeger, P.F., Kohl, S.A., Petersen, J., Maier-Hein, K.H.: NNU-net: a self-configuring method for deep learning-based biomedical image segmentation. Nat. Methods **18**(2), 203–211 (2021). https://doi.org/10.1038/s41592-020-01008-z
10. Kudo, M., et al.: Diagnostic accuracy of imaging for liver cirrhosis compared to histologically proven liver cirrhosis. Intervirology **51**(Suppl. 1), 17–26 (2008). https://doi.org/10.1159/000122595
11. Maier-Hein, L., et al.: Metrics reloaded: Pitfalls and recommendations for image analysis validation. arXiv. org (2206.01653) (2022). https://doi.org/10.1038/s41592-023-02151-z
12. Otsu, N.: A threshold selection method from gray-level histograms. IEEE Trans. Syst. Man Cybern. **9**(1), 62–66 (1979). https://doi.org/10.1109/TSMC.1979.4310076
13. Ramesh, N., Yoo, J.H., Sethi, I.: Thresholding based on histogram approximation. IEE Proc. Vis. Image Signal Proc. **142**(5), 271–279 (1995). https://doi.org/10.1049/ip-vis:19952007
14. Ronneberger, O., Fischer, P., Brox, T.: U-net: convolutional networks for biomedical image segmentation. In: Medical Image Computing and Computer-Assisted Intervention–MICCAI 2015: 18th International Conference, Munich, Germany, October 5-9, 2015, Proceedings, Part III 18, pp. 234–241. Springer (2015). https://doi.org/10.48550/arXiv.1505.04597
15. Rumgay, H., et al.: Global burden of primary liver cancer in 2020 and predictions to 2040. J. Hepatol. **77**(6), 1598–1606 (2022). https://doi.org/10.1016/j.jhep.2022.08.021
16. Sartoris, R., Lazareth, M., Nivolli, A., Dioguardi Burgio, M., Vilgrain, V., Ronot, M.: Ct-based liver surface nodularity for the detection of clinically significant portal hypertension: defining measurement quality criteria. Abdom. Radiol. **45**, 2755–2763 (2020). https://doi.org/10.1007/s00261-020-02519-1
17. Smith, A.D.: Liver surface nodularity quantification from routine CT images as a biomarker for detection and evaluation of cirrhosis. Radiology **280**(3), 771–781 (2016). https://doi.org/10.1148/radiol.2016151542
18. Tian, J., Liu, L., Shi, Z., Xu, F.: Automatic couinaud segmentation from CT volumes on liver using GLC-unet. In: International Workshop on Machine Learning in Medical Imaging, pp. 274–282. Springer (2019). https://doi.org/10.1007/978-3-030-32692-0_32
19. Van der Walt, S., et al.: scikit-image: image processing in python. PEERJ **2**, e453 (2014). https://doi.org/10.7717/peerj.453
20. Yushkevich, P.A., et al.: User-guided 3D active contour segmentation of anatomical structures: significantly improved efficiency and reliability. Neuroimage **31**(3), 1116–1128 (2006). https://doi.org/10.1016/j.neuroimage.2006.01.015

Uncertainty-Aware Deep Learning Classification for MRI-Based Prostate Cancer Detection

Kamilia Taguelmimt[1(✉)], Hong-Phuong Dang[1,2], Gustavo Andrade Miranda[1], Dimitris Visvikis[1], Bernard Malavaud[3], and Julien Bert[1]

[1] LaTIM, UMR1101, INSERM, University of Brest, Brest, France
kamilia.taguelmimt@univ.brest.fr
[2] ECAM Rennes, Louis de Broglie, Bruz, France
[3] Toulouse Oncopole University Cancer Institute, Toulouse, France

Abstract. Early and precise detection of prostate cancer using Magnetic Resonance Imaging (MRI) remains a significant challenge in medical research. Despite the promising potential of Deep Neural Networks (DNNs) for prostate cancer screening, ensuring their reliability is crucial. Accurately quantifying prediction uncertainty in diagnoses is imperative in clinical settings. In this study, we introduce a deep learning model designed not only to detect prostate cancer but also to quantify prediction uncertainty, thus distinguishing between confident and uncertain predictions. Our approach uses a 3D DenseNet-121 backbone for feature extraction and Monte Carlo Dropout (MCD) to approximate Bayesian inference, allowing us to estimate the uncertainty in the model's predictions. We evaluated the model on data from 157 patients, analyzing its reliability and performing an ablation study across different MRI sequences. The model achieved an Area Under the Curve (AUC) of 0.79 across all MRI sequences. In the optimal setup, it classified 75% of predictions as certain and 25% as uncertain, with an AUC of 0.9 for certain predictions. These results clearly demonstrate the model's efficacy in accurately quantifying the reliability of its classifications. By automatically identifying uncertain cases, our approach enables radiologists to focus their attention on these, potentially reducing their workload while enhancing diagnostic accuracy.

Keywords: Prostate Cancer · Biparametric MRI · Deep Learning · Uncertainty

1 Introduction

Prostate cancer (PCa) is the most common cancer among men in 112 countries, accounting for 15% of cancer cases, with projected increases from 1.4 million cases in 2020 to 2.9 million by 2040 worldwide [1]. PCa screening aims to identify cancer as early as possible to improve treatment outcomes. The level of prostate-specific antigen (PSA) can reveal an initial anomaly, indicating a potential cancer. When an anomaly is detected, multiparametric magnetic resonance imaging (mpMRI) is performed. This helps reduce unnecessary biopsies [2] by visually estimating the risk of cancer. If the risk is confirmed, a series of biopsies will be necessary to verify if the lesion is malignant. The visual screening of potential cancer risk is a non-invasive method that combines T2-weighted imaging (T2W), Apparent Diffusion Coefficient (ADC), Diffusion Weighted Imaging (DWI), and dynamic contrast-enhanced MRI. Lesions are classified according to the Prostate Imaging Reporting and Data System (PI-RADS), which enhances the precision of the diagnosis [3]. Although MRI is an excellent screening tool, manual analysis of MRI images is time-consuming and can lead to diagnostic errors. This challenge has driven the development of Computer-aided diagnostic (CAD) tools using machine learning. Several models based on Artificial Intelligence (AI) have been proposed [4]. However, the reliability of these models remains unknown and critical, particularly in the context of cancer risk detection. For clinical applicability, accurately predicting cancer risk with a confidence index is essential.

To address this shortcoming, we propose a deep learning (DL) model that not only detects the presence of PCa but also quantifies prediction uncertainty. This approach enables a second classification, distinguishing certain predictions from uncertain ones, thereby enhancing the applicability of DL-based medical decisions in real clinical contexts. Our main goal is to alleviate the radiologist's workload by focusing his attention only on patients whose automatic diagnosis is uncertain. Moreover, this work presents a comprehensive analysis of the model's reliability, including ablation studies, uncertainty assessment accuracy, confidence matrices, and performance metrics for evaluating uncertainty.

2 Related Work

Examining the advancements in deep learning for PCa classification, we have observed several studies conducted to achieve high performance in diagnostic accuracy. For instance, Liu et al. [5] developed XmasNet, a Convolutional Neural Network (CNN) architecture leveraging 3D MRI data from the PROSTATEx challenge to achieve an AUC of 0.84. Yoo et al. [6] introduced a CNN model to detect cancer from axial DWI images, reaching an AUC of 0.87. Cao et al. [7] proposed FocalNet, a multi-class CNN that detects PCa lesions and predicts their aggressiveness, achieving an AUC of 0.81. Aldoj et al. [8] focused on binary prostate cancer classification using various MRI sequence combinations and a 3D CNN, achieving an AUC of 0.84 for the combination of T2-weighted MRI, ADC, and DWI. Mehrtash et al. [9] developed a CNN model for PCa detection using multiparametric MRI, incorporating zonal information, with an AUC result of 0.80. Lapa et al. [10] developed a recurrent neural network to classify PCa on MRI images, thus providing an improvement in classification performance. More recently, approaches based on zonal segmentation and one-step classification of PCa aggressiveness have emerged. Vente et al. [11] examined 2D MRI through a

U-Net network that detects and classifies lesions simultaneously, incorporating zonal information at different network levels for performance enhancement. The inclusion of zonal information has proven essential for improving detection, and this approach is also supported by several studies that endorse this methodology [12–14].

While significant progress has been made in diagnosing PCa, current methods often do not account for uncertainty. Accurately estimating uncertainty is crucial for improving diagnostic accuracy and ensuring patient safety [15–20]. The proposal represents the first comprehensive effort to evaluate and apply uncertainty estimation in the context of prostate cancer detection. This approach enables the identification of instances where the network demonstrates a lack of confidence in its predictions.

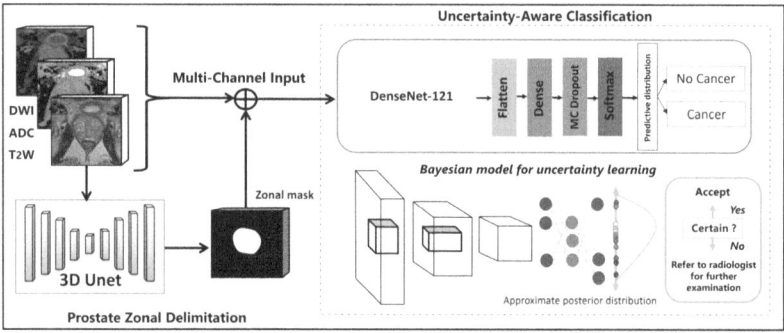

Fig. 1. Schematic diagram of the proposed approach for classification and the associated uncertainty estimation.

3 Methods

3.1 Data Collection and Processing

The data for this study, collected at the *Toulouse Oncopole University Cancer Institute* between 2014 and 2019, include MRI scans (T2W, ADC, DWI) in DICOM format. Patients are divided into two groups: no cancer (78 cases) and cancer (79 cases), with ground truths validated by experts following biopsy results. The MRI images underwent several preprocessing steps using SimpleITK: alignment, cropping, resampling to 96 × 96 × 64, and Min-Max normalization. Prostate segmentation was performed using U-Net [21], trained on the T2W sequence from the PI-CAI database [22], achieving an average Dice Similarity Coefficient (DSC) score of 93.35 during 5-fold cross-validation. The T2W, DWI, ADC sequences, along with the segmentation masks, were combined into a single multichannel image (see Fig. 1). The data was divided into 15% for testing and 85% for training and validation. Data augmentation techniques were applied.

3.2 Uncertainty Modeling with Monte Carlo Dropout

The Bayesian model provides an approach that emphasizes uncertainty, addressing a crucial question "How reliable is our model's prediction?". Traditional Bayesian networks face deployment challenges due to spatial and temporal complexity [23]. A simple way for approximating such networks is to employ Monte Carlo Dropout (MCD) technique [24], which randomly deactivates some neurons during inference, producing multiple stochastic outputs from a single input and enabling an estimation of the model's uncertainty.

To construct our Bayesian network, we adopt a 3D DenseNet-121 backbone as the feature extractor, chosen for its well-known abilities in classification tasks and its capacity to learn rich features while reducing parameters [25]. Subsequently, we introduce an MCD layer after feature extraction to enable Bayesian approximation, resulting in the network named 3D Monte Carlo DenseNet-121 (MC-DenseNet-121). This approach allows our network to generate K predictions from a single testing image. The final prediction per class c is obtained by computing the mean $\mu_p(c)$ over the K predictions as follows:

$$\mu_p(c) = \frac{1}{K}\sum_{k=1}^{K} P_k \qquad (1)$$

where P_k denotes the prediction at iteration k. Uncertainty can then be modeled by computing the predictive standard deviation σ_p and predictive entropy PE, as shown below:

$$\sigma_p = \sqrt{\frac{1}{K}\sum_{k=1}^{K}(P_k - \mu_p)^2} \quad (2) \qquad PE = -\sum_{c} \mu_p(c)\log(\mu_p(c)) \quad (3)$$

σ_p measures the dispersion of predictions around the mean, and PE estimates the uncertainty of the predictions. A lower PE reflects higher confidence in the predictions, whereas a higher PE suggests greater uncertainty.

The complete workflow is as follows: Initially, the 3D MC-DenseNet-121 network takes three MRI sequences (T2W, DWI, ADC) as inputs, along with the segmentation of the prostate gland [14]. Subsequently, the network predicts the presence or absence of cancer using $\mu_p(c)$. These predictions are categorized as correct or incorrect by comparing them with the ground truth values. PE is then used to evaluate the confidence level of the network's predictions by setting a threshold that distinguishes between certain and uncertain predictions. This threshold is determined based on an evaluation outlined in Sect. 3.4. This analysis leads to a secondary classification strategy aimed at objectively categorizing the uncertainty of the predictions into four groups [17,19]:

- True & Certain (TC): number of correct predictions and is certain,
- True & Uncertain (TU): number of incorrect predictions and is uncertain,
- False & Certain (FC): number of incorrect predictions and is certain,
- False & Uncertain (FU): number of correct predictions and is uncertain.

These groups can be seen as an uncertainty confusion matrix, aimed at reducing clinicians' workload by prioritizing cases with prediction uncertainty. Figure 1 illustrates the comprehensive framework for estimating uncertainty.

3.3 Experimental Setup

We compared 3D MC-DenseNet-121 and a standard 3D DenseNet-121 architecture to evaluate the impact of incorporating an MCD layer. Our experiments conducted an ablation study on MRI sequences across three training scenarios: including T2W, ADC, DWI; T2W, ADC only; and solely T2W. This setup allowed us to examine the impact of various MRI sequences on model performance. For each scenario, the mask of the segmented prostate gland was considered. Several dropout rates d_r from 0.1 to 0.9 and multiple inference iterations K from 1 to 50 were used to estimate the uncertainty. We assessed the models' performances using 5-fold cross-validation. The networks were trained using the Adam optimizer and categorical cross-entropy loss function.

3.4 Evaluation Metrics

Initially, we assessed the prostate cancer classification tasks using traditional metrics such as AUC, precision, sensitivity, and specificity. These metrics provide a foundational understanding of the model's performance. For the second categorization, we compute the following specialized metrics based on the uncertainty confusion matrix as follows:

$$U_{\text{Sen}} = \frac{TU}{TU + FC} \quad (4) \qquad U_{\text{Pre}} = \frac{TU}{TU + FU} \quad (6)$$

$$U_{\text{Spe}} = \frac{TC}{TC + FU} \quad (5) \qquad U_{\text{Acc}} = \frac{TU + TC}{TU + TC + FU + FC} \quad (7)$$

where U_{Sen} corresponds to the sensitivity or true positive rate of the traditional confusion matrix. This metric quantifies the model's ability to express its confidence in misclassified samples. U_{Spe} is similar to the specificity metric used in classification. U_{Pre} shares the same concept of precision as used in traditional binary classification. U_{Acc} is calculated similarly to classifier accuracy, taking into account the number of diagonal results compared to the total number of results. These metrics, derived from the uncertainty classification framework, provide insight into the model's capability to reflect confidence in its predictions accurately. By evaluating the model's uncertainty, we can better understand its reliability and identify cases that may require further review or intervention.

4 Results and Discussion

4.1 Classification Performance

For each iteration value K and dropout rates d_r, we assessed the model's performance using all MRI sequences. We identified the model that achieved the highest accuracy (AUC = 0.79), which in our case was obtained with $K = 30$ and $d_r = 0.4$. Due to the stochastic nature of MCD sampling, larger dropout rates provide more variability and may require a larger number of iterations K to stabilize. In order to keep the final computation time required to estimate uncertainty reasonable, the number of iterations K has to be properly chosen. The results for the selected configuration ($K = 30$, $d_r = 0.4$) are shown in Table 1. In this configuration, MC-DenseNet-121 consistently outperforms the standard DenseNet-121, achieving higher AUC values in all MRI sequence scenarios. In the first scenario (T2W, ADC, DWI), MC-DenseNet-121 achieved an AUC of 0.79, with high sensitivity of 0.83 and specificity of 0.75. In the second scenario (T2W and ADC), the AUC increased to 0.83, though sensitivity decreased to 0.75. The last scenario (only T2W) showed slightly lower performance with an AUC of 0.75 and both sensitivity and specificity at 0.75.

Table 1. Detailed Classification Results

Sequence	Model	Sensitivity	Specificity	Precision	AUC
(T2W, ADC, DWI)	DenseNet-121	0.67	**0.83**	**0.80**	0.75
	MC-DenseNet-121	**0.83**	0.75	0.77	**0.79**
(T2W, ADC)	DenseNet-121	0.75	0.75	0.75	0.75
	MC-DenseNet-121	0.75	**0.91**	**0.90**	**0.83**
T2W	DenseNet-121	0.50	0.75	0.67	0.63
	MC-DenseNet-121	**0.75**	0.75	**0.75**	**0.75**

4.2 Uncertainty Estimation Results

The analysis of entropy values in Eq. (3), and predictive distributions for the three models, each using different types of MRI inputs, reveals notable differences in classification performance during the test phase. The distributions in Fig. 2, visualized using the Kernel Density Estimate plot method, illustrate the entropy of correct and incorrect predictions. Low PE indicates high confidence in the predictions, primarily concentrated on the left side of the horizontal axis, while high PE values suggest greater uncertainty, located on the right side. In cases of high PE, it is advisable to refer to the radiologist for further examination. Observation shows that the model with (T2W, ADC, DWI) shows greater confidence in its predictions. This qualitative approach highlights the model's ability to assess and express its confidence in the predictions it generates. The determination of the second classification (certain and uncertain) was based using a threshold on the entropy value, which ranges from 0 to 1 and defines the acceptable variation around a prediction. For each threshold value, metrics from the uncertainty confusion matrix (Sect. 3.4) were calculated and displayed in Fig. 3. For example, considering PCa detection a good compromise is a setup using T2W+ADC+DWI and a PE threshold $= 0.2$. In this case 75% of the predictions was classified as certain and 25% as uncertain, leading to an AUC of 0.9 for certain predictions. However, 8% of the test set was classified as certain and incorrect, which is the worse case for cancer detection. It is also important to note that the number of correctly classified images is much higher than the number of misclassified images. Metrics based on the uncertainty confusion matrix according to the entropy threshold (U_{Sen}, U_{Spe}, U_{Acc} and U_{Pre}) are presented in Fig. 4. Results indicate that all three models are quite capable of flagging incorrect predictions. The values of U_{Pre} are smaller than those of other metrics, due to the fact that the number of FU predictions is much smaller than the number of TU predictions. The values of U_{Acc} and U_{Spe} for the three models are at acceptable reliability levels. Detailed values of the metrics confusion matrix for a PE threshold value of 0.2 are shown in Table 2. The model incorporating all MRI sequences was significantly the most reliable, with a U_{Acc} of 0.79.

Fig. 2. Distribution of predictive entropy (PE) for correct and incorrect predictions across different MRI sequence combinations. From left to right: the PE for the model with T2W MRI, followed by the PE for the model with T2W and ADC sequences, and the PE for the model with MRI sequences (T2W, ADC, DWI).

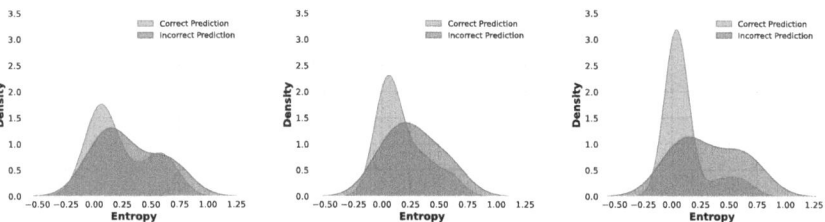

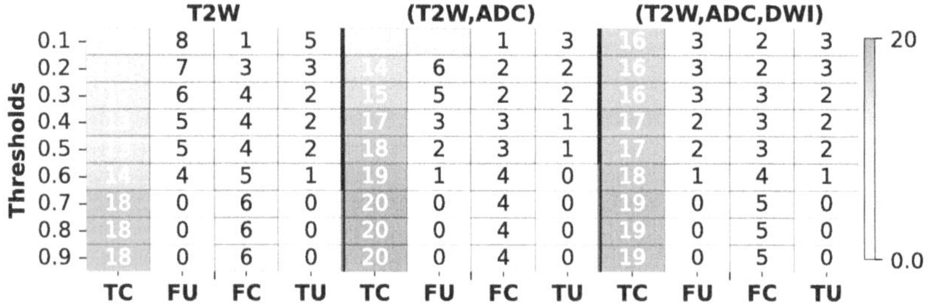

Fig. 3. The uncertainty confusion matrix results for thresholds ranging from 0.1 to 0.9, with a step of 0.1, for the three comparative models.

The ablation study and the results of the second classification have shown the importance of using all three MRI sequences, highlighting that DWI and ADC provide significant information for reliable cancer classification.

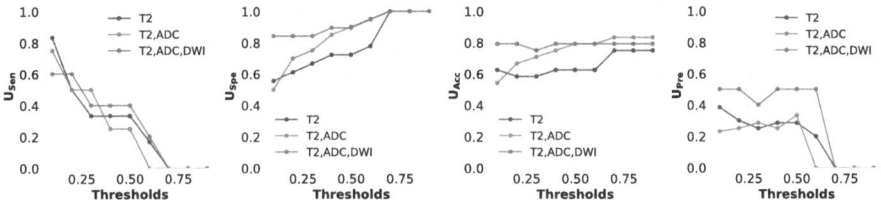

Fig. 4. Quantitative evaluation of different modality combinations (T2W, ADC, DWI) as network inputs. Uncertainty accuracy, sensitivity, specificity, and precision are calculated for threshold values between 0.1 and 0.9 with a step of 0.1.

Table 2. Detailed Classification Results - Uncertainty (Thres = 0.2)

Model	Metric	T2W+ADC+DWI	T2W+ADC	T2W
Uncertainty (Thres = 0.2)	U_{Sen}	0.60	0.50	0.50
	U_{Pre}	0.50	0.25	0.30
	U_{Spe}	0.84	0.70	0.61
	U_{Acc}	**0.79**	0.67	0.58

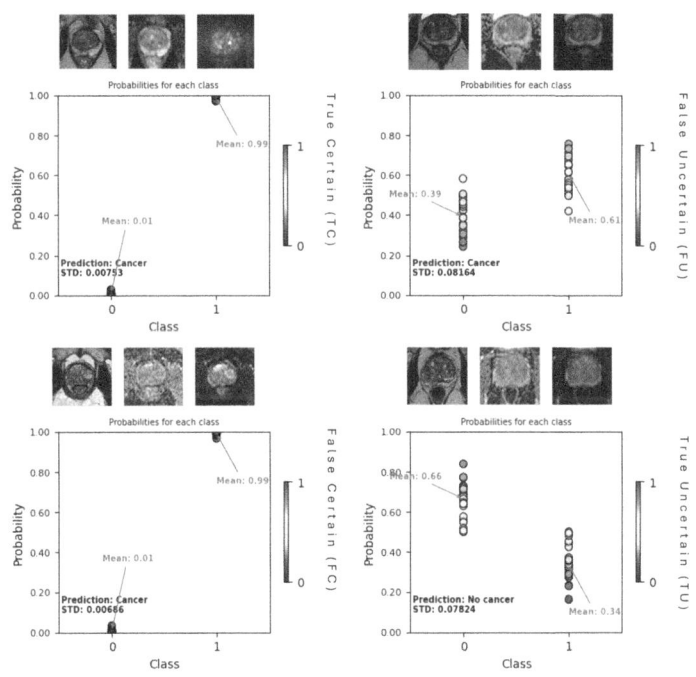

Fig. 5. Example of test results obtained from the MC-DenseNet-121 model with all MRI sequences (T2W, ADC, DWI) using 30 iterations. Four patients are presented, with their TC, FU, FC and TU obtained with the 0.2 threshold, as well as their predictions for belonging to class 0 (no cancer) or 1 (cancer).

Figure 5 showcased four patients from different classes, displaying their corresponding predictive distributions and the resulting uncertainty labels. Among these examples, one of them (False and Certain) show a critical case where the model predicts with high certainty an incorrect classification. The standard deviation and the PE were equivalent in terms of values to the example of the True and Certain case, making it impossible to distinguish certain from uncertain. Further investigation needs to explore to understand these two uncommon cases.

5 Conclusion

Our study explored enhancing the diagnosis of prostate cancer by integrating the Monte-Carlo Dropout technique with the 3D DenseNet-121 model applied to MRI scans. The results demonstrate that this approach not only improves the model's accuracy but also enables a more reliable assessment of prediction uncertainty. Understanding uncertainty in depth is crucial for enhancing patient safety, providing more accurate diagnoses, and avoiding potential errors. Our future work will involve studying our approach on a diverse and larger clinical population. The aim will be to understand and quantify how the number of False Certain predictions can be reduced, especially by exploring other types of feature extraction architectures. In prostate cancer screening, this is a critical aspect. However, it is also important to minimize the number of uncertain predictions (both False and Correct), as these cases will still require examination by radiologists despite the diagnostic tool.

Acknowledgments. This work was partly supported by the French State under the Future Investments Program through the DIANA project (PSPC-DIANA-2021 BPI).

References

1. James, N.D., et al.: The Lancet Commission on prostate cancer: planning for the surge in cases. Lancet **403**(10437), 1683–1722 (2024)
2. Fütterer, J.J., et al.: Can clinically significant prostate cancer be detected with multiparametric magnetic resonance imaging? A systematic review of the literature. Eur. Urol. **68**(6), 1045–1053 (2015)
3. Turkbey, B., et al.: Prostate imaging reporting and data system version 2.1: update of prostate imaging reporting and data system version 2. Eur. Urol. **76**(3), 340 (2019)
4. Cuocolo, R., et al.: Machine learning applications in prostate cancer magnetic resonance imaging. Eur. Radiol. Exp. **3**(1), 1–8 (2019)
5. Liu, S., Zheng, H., Feng, Y., Li, W.: Prostate cancer diagnosis using deep learning with 3D multiparametric MRI. SPIE Med. Imaging Comput.-Aid. Diag. **10134**, 581–584 (2017)
6. Yoo, S., Gujrathi, I., Haider, M.A., Khalvati, F.: Prostate cancer detection using deep convolutional neural networks. Sci. Rep. **9**(1), 1–10 (2019)
7. Cao, R., et al.: Joint prostate cancer detection and Gleason score prediction in mp-MRI via FocalNet. IEEE Trans. Med. Imaging **38**(11), 2496–2506 (2019)

8. Aldoj, N., Lukas, S., Dewey, M., Penzkofer, T.: Semi-automatic classification of prostate cancer on multi-parametric MR imaging using a multi-channel 3D convolutional neural network. Eur. Radiol. **30**(2), 1243–1253 (2020)
9. Mehrtash, A., et al.: Classification of clinical significance of MRI prostate findings using 3D convolutional neural networks. In: SPIE Medical Imaging 2017: Computer-Aided Diagnosis, vol. 10134, pp. 589–592 (2017)
10. Lapa, P., Castelli, M., Gonçalves, I., Sala, E., Rundo, L.: A hybrid end-to-end approach integrating conditional random fields into CNNs for prostate cancer detection on MRI. Appl. Sci. **10**(1), 338 (2020)
11. de Vente, C., Vos, P., Hosseinzadeh, M., Pluim, J., Veta, M.: Deep learning regression for prostate cancer detection and grading in bi-parametric MRI. IEEE Trans. Biomed. Eng. **68**(2), 374–383 (2021)
12. Pellicer-Valero, O.J., et al.: Deep Learning for fully automatic detection, segmentation, and Gleason Grade estimation of prostate cancer in multiparametric Magnetic Resonance Images. Sci. Rep. **12**(1), 1–13 (2022)
13. Duran, A., Dussert, G., Rouvière, O., Jaouen, T., Jodoin, P.M., Lartizien, C.: ProstAttention-Net: a deep attention model for prostate cancer segmentation by aggressiveness in MRI scans. Med. Image Anal. **77**, 102347 (2022)
14. Saha, A., Hosseinzadeh, M., Huisman, H.: End-to-end prostate cancer detection in bpMRI via 3D CNNs: effects of attention mechanisms, clinical priori and decoupled false positive reduction. Med. Image Anal. **73**, 102155 (2021)
15. Kurz, A., et al.: Uncertainty estimation in medical image classification: systematic review. JMIR Med. Inform. **10**(8), e36427 (2022)
16. Gour, M., Jain, S.: Uncertainty-aware convolutional neural network for COVID-19 X-ray images classification. Comput. Biol. Med. **140**, 105047 (2021)
17. Asgharnezhad, H., et al.: Objective evaluation of deep uncertainty predictions for COVID-19 detection. Sci. Rep. **12**(1), 815 (2022)
18. Song, B., et al.: Bayesian deep learning for reliable oral cancer image classification. Biomed. Opt. Express **12**, 6422–6430 (2021)
19. Hamedani-KarAzmoudehFar, F., et al.: Breast cancer classification by a new approach to assessing deep neural network-based uncertainty quantification methods. Biomed. Signal Process. Control **79**, 104057 (2023)
20. Mobiny, A., Singh, A., Van Nguyen, H.: Risk-aware machine learning classifier for skin lesion diagnosis. J. Clin. Med. **8**(8), 1241 (2019)
21. Ronneberger, O., Fischer, P., Brox, T.: U-Net: convolutional networks for biomedical image segmentation. In: Navab, N., Hornegger, J., Wells, W.M., Frangi, A.F. (eds.) MICCAI 2015. LNCS, vol. 9351, pp. 234–241. Springer, Cham (2015). https://doi.org/10.1007/978-3-319-24574-4_28
22. Saha, A., et al.: Artificial Intelligence and Radiologists at Prostate Cancer Detection in MRI: The PI CAI Challenge. Abstract from RSNA 2022, Chicago, United States (2022)
23. Ghahramani, Z.: Probabilistic machine learning and artificial intelligence. Nature **521**(7553), 452–459 (2015)
24. Gal, Y., Ghahramani, Z.: Dropout as a Bayesian approximation: representing model uncertainty in deep learning. In: International Conference on Machine Learning, pp. 1050–1059. PMLR (2016)
25. Huang, G., Liu, Z., Van Der Maaten, L., Weinberger, K. Q.: Densely connected convolutional networks. In: Proceedings of the IEEE Conference on Computer Vision and Pattern Recognition, pp. 4700–4708 (2017)

Generalized Polyp Detection from Colonoscopy Frames Using Proposed EDF-YOLO8 Network

Alyaa Amer[1,4], Alaa Hussein[2], Noushin Ahmadvand[3], Sahar Magdy[3], Abas Abdi[3], Nasim Dadashi Serej[3], Noha Ghatwary[1(✉)], and Neda Azarmehr[3]

[1] Arab Academy for Science and Technology, Smart Village Campus, Cairo, Egypt
noha.ghatwary@aast.edu
[2] Pharos University in Alexandria, Alexandria, Egypt
[3] School of Computing and Engineering, University of West London, London, UK
[4] American University in the Emirates, Dubai, UAE

Abstract. Colon cancer is among the leading causes of cancer-related death worldwide for both men and women, with colorectal polyps serving as a significant predisposing factor. Early polyp identification and removal-the precursors to colorectal cancer-is essential to its prevention. Colonoscopy is considered the gold standard for colorectal cancer screening because it allows for the immediate removal of polyps, preventing them from developing into cancer. Despite its effectiveness, conventional colonoscopy is time-consuming, highly labor-intensive, and prone to human mistakes. Therefore, we modified the efficient object detection model, YOLO-V8, to develop our novel approach, EDF-YOLO8, for automating polyp identification. Our model employs deformable convolution in the bottleneck as a robust solution for effectively detecting polyps of various sizes. We enhance the effectiveness of our model by incorporating the Exponential Linear Unit (ELU), which further increases the detection accuracy and tends to accelerate the model learning process. We trained and tested the suggested model on two distinct datasets from publicly accessible sources and conducted thorough assessments to ensure its robustness and generalizability. The proposed model achieved an outstanding performance, attaining a mAP50 score of 0.931 and 0.894 for the Kvasir and Polypgen datasets, respectively. Performance analysis demonstrates the efficiency and robustness of our model in accurately detecting polyps from colonoscopic frames from different datasets.

Keywords: Colorectal cancer · Polyp detection · YOLOv8

1 Introduction

Colorectal cancer (CRC) is the third most common cancer globally, accounting for approximately 10% of all cancer cases. It is also the second leading cause of cancer-related deaths worldwide [1]. Colonoscopy is considered the gold standard in CRC screening because it immediately removes polyps that develop in the mucus layer of the colon or rectum and are often noncancerous growths at first but lead to CRC over time [2]. Detection and removal of polyps are therefore important to prevent the development of colorectal cancer. Despite the promising results in polyp detection, the procedure is time-consuming and prone to human errors. Depending on the size of the polyp, there is a 15–30% chance of missing polyps during back-to-back colonoscopies [3]. Therefore, there remains a need to develop a computer-aided system to ease the examination process and help endoscopists improve the detection rate of polyps.

Most existing advanced algorithms for polyp detection utilize Convolutional Neural Networks (CNNs) [4,5]. One of the major challenges for CNN-based polyp detection is the multi-scale problem caused by the diverse sizes and shapes of polyps, and the camera movements during examination exacerbate this issue, making it difficult for CNN models to effectively extract features and match detection boxes accurately. Moreover, studies show that the polyp detection ability diminishes as the size of the polyps decreases [6], and even a marginal increase of 1% in the polyp detection recall rate can reduce the risk of colon cancer by 3% [7].

Recently, high-performance object detection algorithms such as You Only Look Once (YOLO) [8] were applied to polyp detection and showed significant promise in enhancing the sensitivity and specificity of polyp detection systems. Therefore, in this study, we proposed enhancing the YOLO-V8 network with several key modifications to improve its performance and robustness. One significant enhancement involves the integration of deformable convolutions [9]. Deformable convolutions (DC) dynamically adjust their sampling grid based on input features, enabling the model to more effectively capture and represent the diverse shapes and sizes of polyps, thereby improving detection rates. This adaptability enables the convolutional kernel to adjust its sampling locations, enhancing the detection of small and irregularly shaped polyps that standard CNN approaches might miss. Additionally, we incorporated the ELU activation function to further enhance the model's performance, mitigate the vanishing gradient problem, and facilitate the training process. These enhancements make our model more effective at detecting small and irregularly shaped polyps, which are often missed by other CNN approaches.

2 Background and Related Literature

In a systematic review of the application of artificial intelligence for polyp detection during colonoscopy detailed in [17], researchers compared the disparities between colonoscopy procedures with and without AI. Their investigation revealed that an AI-driven polyp detection system can notably improve the detection rate of non-advanced adenomas and smaller polyps during colonoscopy.

However, developing a colorectal polyp detection model for analyzing colonoscopies presents several challenges. These challenges include (1) a lack of diverse samples to cover the wide range of possible polyp variations [12]; (2) improper patient examination, which can lead to misidentifications, with the model potentially mistaking non-polyp objects such as feces or normal blood vessels for polyps [13,14]; (3) the demand for high computational resources [16], and optimizing detection speed often results in decreased performance and increased inaccurate predictions; (4) the reflection of white light in colonoscopies, which can either obscure polyps or create polyp-like anomalies, leading to poor performance [12].

Recently, YOLO has been applied to polyp detection, resulting in more successful outcomes. YOLO models are relatively superior to their competitors in real-time detection as they integrate the Region Proposal Network (RPN) and classification stage into a single network. Liu et al. [18] proposed a YOLOv3-based approach that integrates spatiotemporal information into a two-dimensional CNN-based real-time object detection network, aiming to reduce the high rate of missed polyp detections and false positives. This method achieved an F1 score of 77%. Pacal and Karaboga [19] introduced a real-time automatic polyp detection system by adapting the YOLOv4 algorithm. They enhanced the architecture by integrating the cspnet network, employing the mish activation function, Diou loss function, and incorporating transformer blocks, resulting in improved performance and speed. Durak et al. [20] conducted training on YOLO-V4, CenterNet, EfficientDet, and YOLO-V3, for automatic gastric polyp detection. Their experimental results highlighted that YOLO-V4 demonstrated superior performance compared to other methods, demonstrating its effectiveness for integration into computer-aided detection systems for automatic polyp detection. Additionally, Ahmet Karaman et al. [21] introduced a novel integration of the ABC algorithm [19] with YOLO-based algorithms. This integration optimizes operations more efficiently in a single process, leading to time and hardware cost savings.

According to the reviewed literature in [15], the YOLO algorithm is the most widely used method for detecting polyps in static images. Advancements in the YOLO series show promise for more efficient and accurate polyp detection during colonoscopy. YOLO's popularity among researchers can be attributed to its ability to detect objects faster than other detection algorithms, making it ideal for real-time polyp detection tasks. Specifically, YOLO-V8 has achieved state-of-the-art performance through optimizations in model structure, anchor box or anchor-free schemes, and the implementation of diverse data augmentation techniques, significantly enhancing overall detection precision compared to its predecessors. Additionally, YOLO-V8 is optimized to run smoothly on standard hardware, making it a practical choice for real-time object detection tasks. Thereby, we compared the performance of YOLO-V8 with two state-of-the-art methods in Sect. 4, and it outperformed them both. As a result, YOLO-V8 has been selected as the base model for our work.

3 Methods

The main framework of our proposed model "EDF-Yolo" is presented in Fig. 1. Inspired by recent advancements in neural network design. We incorporated two key modifications to enhance the model's performance and robustness. Originally, the YOLOv8 network included a CBS module comprised of Convolution, Batch Normalization, and SiLu activation functions. Figure 2a illustrates our model's Conv_ELu block, which employs ELU (Exponential Linear Unit) activation functions rather than the conventional SiLU (Sigmoid Linear Unit) activation functions. ELU is known for its robustness in handling vanishing gradient problems and providing a faster and smoother convergence, improving the model's performance in detecting subtle features of polyps.

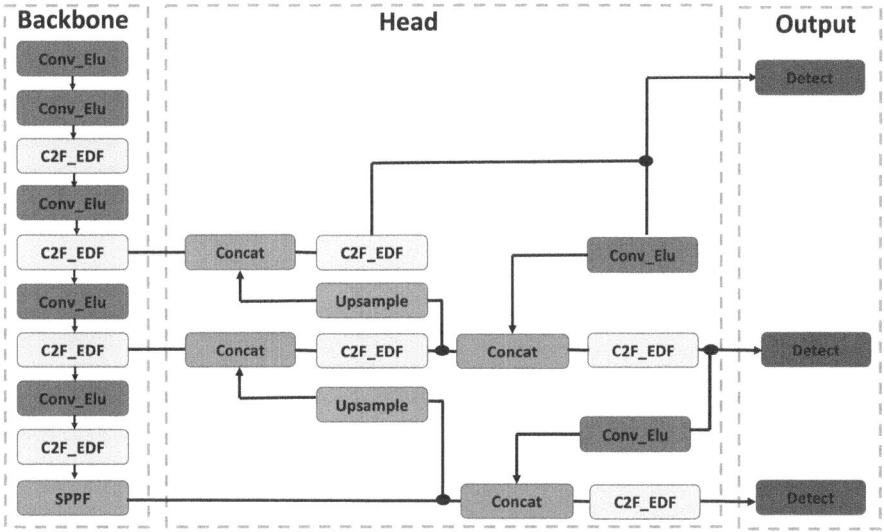

Fig. 1. Proposed EDF-YOLO8 with updated blocks *Conv_Elu* and *C2F_EDF*

Additionally, we introduce deformable convolution layers in the Cross Stage Partial Network with the Fusion (C2F) module of the Neck layer through a block named C2F_EDF Block as presented in Fig. 2b. The neck layer in the YOLO architecture serves as an intermediary between the backbone and the head, refining and enhancing the features extracted by the backbone. Deformable convolutions in the C2F module enable the network to adaptively adjust its receptive fields for targeting multi-scale objects. This increases the model's flexibility in capturing the irregular shapes and varying sizes of polyps, which are common challenges in medical image analysis.

For the utmost confidence in the robustness and generalizability of our proposed model, we have thoroughly trained and tested it using the Kvasir and PolypGen datasets. Both datasets are widely recognized for their diversity and quality. Furthermore, we have trained the model using the Kvasir dataset and tested it on the PolypGen dataset, and vice versa. This is to thoroughly evaluate the model's ability to generalize across different data sources. This meticulous cross-dataset validation approach ensures that our model performs reliably in diverse real-world settings.

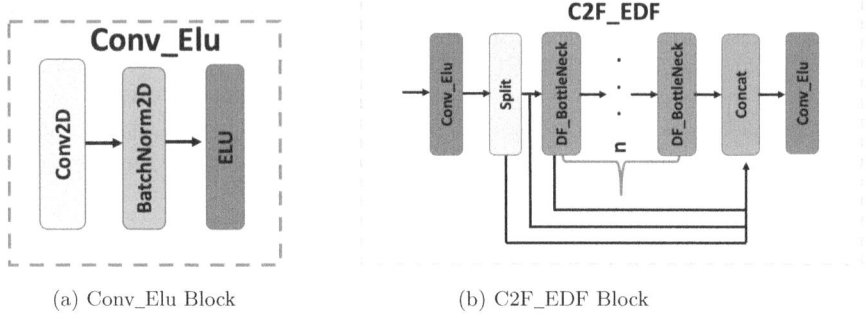

Fig. 2. The detailed modification inside each block where (a) represents the 'Conv_Elu' after replacing activation function and (b) represents the 'C2F_EDF' after incorporating the Deformable Convolution

4 Experimental Setup and Results

4.1 Dataset

This study used two publicly available and diverse datasets of colonoscopy images: Kvasir and PolypGen. These datasets contain a wide variety of polyp types and sizes. The Kvasir dataset [10] has 1,001 images, while the PolypGen dataset [11] consists of 5,055 images, including both single and sequence frames, and is particularly useful for detecting polyps in different contexts and motions, enhancing the robustness of the detection algorithm. To train deep learning models, we split both the Kvasir and PolypGen datasets into 60% for training, 20% for validation, and 20% for testing to ensure a balanced evaluation of the model's performance across different datasets. The weights are initialized randomly with a gaussian distribution ($\mu = 0$, $\sigma = 0.01$). The initial learning rate is set to ($1e - 4$) and trained for 800 epochs.

4.2 Results and Discussion

Table 1 presents a quantitative comparison of polyp detection performance using baseline models (several versions of YOLOv8, YOLOv9 and DETR) on the Kvasir dataset. The results demonstrate that YOLOv8x outperformed other versions with a precision of 0.932 and mAP50 of 0.933. YOLOv8s also showed competitive performance, with a recall of 0.886 and mAP50-90 of 0.762. However, DETR demonstrated poor performance in accurately detecting polyps within the Kvasir dataset compared to the YOLOv9 and YOLOv8 variants. Moreover, we individually assess each contribution phase to assess the EDF-YOLO8. The results of each step are depicted separately in Table 2 when evaluating the lowest-performing version, YOLOv8n. We also conducted the McNemar statistical test to confirm the significance of the detection performance, as shown in Table 2, comparing the model with and without the enhancements. The test revealed that the findings were significantly different at the 5% level (p-value < 0.05).

Table 1. The detection results of the baseline models tested on the Kvasir dataset. The metrics with the highest scores are indicated in bold.

	Precision	Recall	mAP50	mAP50-90
YOLOv8n	0.894	0.832	0.907	0.731
YOLOv8s	0.894	**0.886**	0.909	**0.762**
YOLOv8x	**0.932**	0.833	**0.933**	0.749
YOLOv9c	0.876	0.852	0.917	0.761
DETR	0.708	0.783	0.889	0.708

The proposed EDF-YOLO8s, which incorporates deformable convolutional layers and ELU activations, demonstrates improvements in detection performance across both the Kvasir and PolypGen datasets, as shown in Table 3. On the Kvasir dataset, EDF-YOLO8s achieved the highest precision (0.940) and recall (0.862), with a mAP50 of 0.931. EDF-YOLO8n also showed good performance with mAP50-90 of 0.769. On the PolypGen dataset, EDF-YOLO8n achieved the highest precision (0.939) and mAP50 (0.894). EDF-YOLO8s had a recall of 0.822 and a mAP50-90 of 0.724, indicating comparable performance. These findings illustrate the robustness and adaptability of the EDF-YOLO8s architecture across Kvasir and PolypGen datasets. Additionally, the effectiveness of the EDF-YOLO8s model is evident in the sample images shown in Fig. 3. The bounding boxes and confidence scores highlight its identification accuracy across various polyp sizes and varying conditions. Furthermore, the model managed not to detect false polyps from negative samples in the PolypGen dataset, as presented in the last two images.

To further emphasize the robustness and generalizability of the EDF-YOLO8s and EDF-YOLO8n models, Table 4 investigates their performance when trained on one dataset and tested on another. It is noted that both models provide good and comparable performance when trained on Kvasir and tested on PolypGen, and these results are fairly close to those when trained on PolyGen and tested on Kvasir. Notably, a higher performance is achieved from both when trained on the more diverse and larger dataset, PolypGen. Overall, the precision results showcase the generalization capability of the models, indicating their reliability and robustness in diverse real-world scenarios.

Table 2. Ablation experiments results tested on the Yolo8n tested on both the Kvasir and Polypgen datasets.

	Precision	Recall	mAP50	mAP50-90
Kvasir Dataset				
YOLOv8n	0.894	0.832	0.907	0.731
YOLOv8n + Conv_ELU	0.904	0.835	0.909	0.737
YOLOv8n + C2F_EDF	0.914	0.842	0.912	0.751
EDF-YOLO8n	**0.923**	**0.858**	**0.919**	**0.769**
PolypGen Dataset				
YOLOv8n	0.891	0.794	0.874	0.697
YOLOv8n + Conv_ELU	0.904	0.799	0.879	0.706
YOLOv8n + C2F_EDF	0.919	0.809	0.885	0.707
EDF-YOLO8n	**0.939**	0.810	**0.894**	**0.717**

Table 3. The detection results for the proposed EDF-YOLO8 on the Kvasir and Polypgen datasets. The metrics with the highest scores are indicated in bold.

	Precision	Recall	mAP50	mAP50-90
Kvasir Dataset				
EDF-YOLO8n	0.923	0.858	0.919	**0.769**
EDF-YOLO8s	**0.940**	**0.862**	**0.931**	0.768
EDF-YOLO8x	0.865	0.853	0.901	0.734
PolypGen Dataset				
EDF-YOLO8n	**0.939**	0.810	**0.894**	0.717
EDF-YOLO8s	0.917	**0.822**	0.888	**0.724**
EDF-YOLO8x	0.915	0.821	0.885	0.706

Table 4. An assessment of the generalizability of the model across different training and testing datasets.

Model	Train Data	Test Data	Precision	Recall	mAP50	mAP50-90
EDF-YOLO8n	Kvasir Dataset	PolypGen Dataset	0.737	0.527	0.591	0.444
	PolypGen Dataset	Kvasir Dataset	0.861	0.810	0.876	0.684
EDF-YOLO8s	Kvasir Dataset	PolypGen Dataset	0.671	0.562	0.588	0.431
	PolypGen Dataset	Kvasir Dataset	0.875	0.786	0.884	0.687

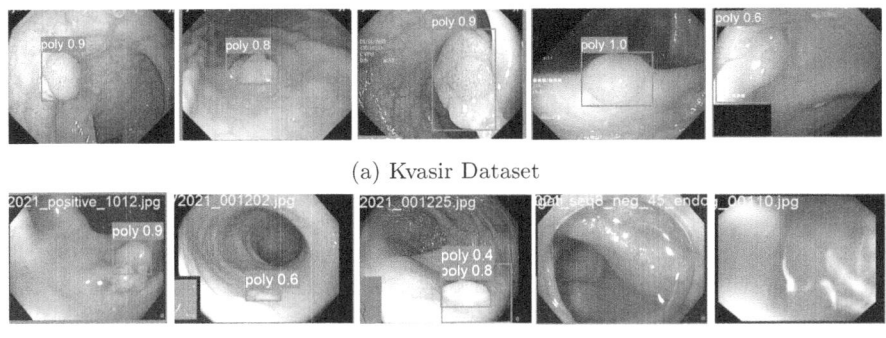

(a) Kvasir Dataset

(b) PolypGen Dataset

Fig. 3. Sample of the detection results from both Kvasir and PolypGen datasets

5 Conclusion

This paper proposes a novel EDF-YOLO8 that automatically detects polyps from colonoscopy images. We trained and evaluated the various model architectures available in the YOLOv family (YOLOv8n, YOLOv8s, YOLOv8x, and YOLOv9c) on two different datasets, the Kvasir and PolypGen datasets. The EDF-YOLO8 models were better at finding polyps than regular YOLOv8 networks after they were enhanced with ELU activations and deformable convolutions. The outcomes show that the suggested detection approach successfully identifies polyps of various sizes. Notably, models trained on the more diverse and generalised PolypGen dataset performed better when tested on Kvasir. To assess their generalizability, future research will test the models and add more colonoscopy polyp datasets.

References

1. Sung, H., et al.: Global cancer statistics 2020: GLOBOCAN estimates of incidence and mortality worldwide for 36 cancers in 185 countries. CA Cancer J. Clin. **71**(3), 209–249 (2021)
2. Sawicki, T., Ruszkowska, M., Danielewicz, A., Niedźwiedzka, E., Arłukowicz, T., Przybyłowicz, K.E.: A review of colorectal cancer in terms of epidemiology, risk factors, development, symptoms and diagnosis. Cancers **13**(9), 2025 (2021)
3. Matsuda, T., Ono, A., Kakugawa, Y., Matsumoto, M., Saito, Y.: Impact of screening colonoscopy on outcomes in colorectal cancer. Jpn. J. Clin. Oncol. **45**(10), 900–905 (2015)
4. Pacal, I., Karaboga, D., Basturk, A., Akay, B., Nalbantoglu, U.: A comprehensive review of deep learning in colon cancer. Comput. Biol. Med. **126**, 104003 (2020)
5. Rahman, M.M., Wadud, M.A.H., Hasan, M.M.: Computerized classification of gastrointestinal polyps using stacking ensemble of convolutional neural network. Inform. Med. Unlocked **24**, 100603 (2021)
6. Cao, C., Wang, R., Yu, Y., Zhang, H., Yu, Y., Sun, C.: Gastric polyp detection in gastroscopic images using deep neural network. PLoS ONE **16**(4), e0250632 (2021)

7. Corley, D.A., et al.: Adenoma detection rate and risk of colorectal cancer and death. N. Engl. J. Med. **370**(14), 1298–1306 (2014)
8. Redmon, J., Divvala, S., Girshick, R., Farhadi, A.: You only look once: unified, real-time object detection. In: Proceedings of the IEEE Conference on Computer Vision and Pattern Recognition, pp. 779–788 (2016)
9. Dai, J., et al.: Deformable convolutional networks. In: Proceedings of the IEEE International Conference on Computer Vision, pp. 764–773 (2017)
10. Pogorelov, K., et al.: Kvasir: a multi-class image dataset for computer aided gastrointestinal disease detection. In: Proceedings of the 8th ACM on Multimedia Systems Conference, pp. 164–169 (2017)
11. Ali, S., et al.: A multi-centre polyp detection and segmentation dataset for generalisability assessment. Sci. Data **10**(1), 75 (2023)
12. Nisha, J., Gopi, V., Palanisamy, P.: Automated colorectal polyp detection based on image enhancement and dual-path CNN architecture. Biomed. Sig. Process. Control **73**, 103465 (2022)
13. Yang, K.: Automatic polyp detection and segmentation using shuffle efficient channel attention network. Alex. Eng. J. **61**, 917–926 (2022)
14. Gong, E., et al.: No-code platform-based deep-learning models for prediction of colorectal polyp histology from white-light endoscopy images: development and performance verification. J. Pers. Med. **12**, 963 (2022)
15. ELKarazle, K., Raman, V., Then, P., Chua, C.: Detection of colorectal polyps from colonoscopy using machine learning: a survey on modern techniques. Sensors **23**, 1225 (2023)
16. Puyal, J., et al.: Polyp detection on video colonoscopy using a hybrid 2D/3D CNN. Med. Image Anal. **82**, 102625 (2022)
17. Barua, I.: Artificial intelligence for polyp detection during colonoscopy: a systematic review and meta-analysis. Endoscopy **53**, 277–284 (2021)
18. Liu, X., Guo, X., Liu, Y., Yuan, Y.: Consolidated domain adaptive detection and localization framework for cross-device colonoscopic images. Med. Image Anal. **71**, 102052 (2021)
19. Pacal, I., Karaboga, D.: A robust real-time deep learning based automatic polyp detection system. Comput. Biol. Med. **134**, 104519 (2021)
20. Durak, S., et al.: Deep neural network approaches for detecting gastric polyps in endoscopic images. Med. Biol. Eng. Comput. **59**, 1563–1574 (2021)
21. Karaman, A., et al.: Hyper-parameter optimization of deep learning architectures using artificial bee colony (ABC) algorithm for high performance real-time automatic colorectal cancer (CRC) polyp detection. Appl. Intell. **53**, 15603–15620 (2022)
22. Pacal, I., Karaboga, D.: A robust real-time deep learning based automatic polyp detection system. Comput. Biol. Med. **134**, 104519 (2021)

AI-Assisted Laryngeal Examination System

Chiara Baldini[1,2(✉)], Muhammad Adeel Azam[1,2], Madelaine Thorniley[1], Claudio Sampieri[3,6], Alessandro Ioppi[4,6], Giorgio Peretti[5,6], and Leonardo S. Mattos[1]

[1] Biomedical Robotics Laboratory, Department of Advanced Robotics, Istituto Italiano di Tecnologia, Genoa, Italy
{chiara.baldini,leonardo.demattos}@iit.it
[2] Department of Informatics, Bioengineering, Robotics, Systems Engineering, University of Genova, Genoa, Italy
[3] Department of Experimental Medicine, University of Genoa, Genova, Italy
[4] Department of Otolaryngology, Hospital Clínic, Barcelona, Spain
[5] Department of Otorhinolaryngology-Head and Neck Surgery, S. Chiara Hospital, Trento, Italy
[6] Unit of Otorhinolaryngology-Head and Neck Surgery, IRCCS Ospedale Policlinico San Martino, Genoa, Italy

Abstract. Laryngeal cancer (LC) and other benign conditions are major concerns in modern ear, nose, and throat medicine. A comprehensive evaluation of the larynx should employ flexible or rigid endoscopes to identify early-stage lesions, possibly enhanced with advanced imaging techniques such as Narrow Band Imaging (NBI) to empower tissue visualization. Factors that make the detection, diagnosis, and treatment of LC challenging include the huge amount of uninformative frames and the expertise-dependent nature of the assessment, leading to time-consuming procedures with high cognitive loads and the possibility of missed detections and misdiagnoses, especially for less-experienced clinicians. Deep Learning (DL) approaches have recently been studied regarding frame quality assessment, abnormal mass identification, and their margins definition to improve diagnostic accuracy and surgical outcomes. In this work, we proposed the integration of several Convolutional Neural Networks (CNNs) into a single computer-aided system for the assistance of less-experienced otolaryngologists, by directing their attention toward good-quality frames from which lesions can be automatically detected and characterized. We addressed the following challenging tasks: informative frame selection, lesion detection, lesion classification, and lesion segmentation. The developed system demonstrated a good trade-off between efficacy metrics and real-time performance, and the potential for clinical applications.

Keywords: Laryngeal lesions · Endoscopy · CADe/x system

1 Introduction

1.1 Laryngeal Cancer and Clinical Challenges

Laryngeal Cancer (LC) is a primary concern in modern oncology and ear, nose, and throat medicine, accounting for a significant percentage of head and neck malignancies worldwide [1]. Additionally, the larynx can be affected by various other disorders, including contact ulcers, laryngitis, benign laryngeal tumors, and paralysis of the vocal cord. Since many of these conditions cause changes in voice quality, dysphonia should never be underestimated and should be regarded as a common symptom in their early stages. However, the diverse nature of laryngeal disorders, each varying significantly from the others, highlights the crucial need for comprehensive prompt diagnostic and therapeutic methods to enable good prognosis and enhance clinical outcomes. Indeed, current clinical practice focuses on a multidisciplinary approach incorporating flexible and/or rigid endoscopy as a pivotal component [2]. A suitable differential diagnosis should encompass thorough endoscopic evaluation of the larynx, advanced imaging modalities such as Narrow Band Imaging (NBI) to enhance visualization and characterization of possible laryngeal lesions, and biopsy-based histopathological confirmation of suspected malignancies. This comprehensive strategy aims to improve diagnostic accuracy and provide tailored treatment plans for patients, avoiding the use of further diagnostic imaging techniques such as CT or MRI. Treatment typically implies a delicate balance between surgical intervention, radiotherapy, and chemotherapy, tailored to disease stage, patient factors, and functional preservation goals.

Otolaryngologists can encounter several challenges in the diagnostic and treatment pathway of LC. In the first phase, which involves outpatient examination with flexible endoscopy for laryngeal initial investigation, challenges deal with the low informative content of endoscopic videos due to noise, blur, and occlusions. This leads to time-consuming procedures that impose a high cognitive load on the clinician, who needs to carefully look at the recordings to detect signs of disease. In addition, factors such as otolaryngologists' expertise and equipment quality significantly impact the precision of the assessment. For instance, although NBI helps enhance outcomes, its adoption requires extensive training, making the diagnostic process highly operator-dependent, and resource constraints may limit its availability. These problems are accentuated for less-experienced clinicians, leading to missed detections and misdiagnoses. Failure to identify or classify malignant lesions results in a misperception of the problem, with adverse consequences for patients. Hence, biopsy is often required to validate the assumed diagnosis, increasing time, costs, and risks for the patients. Finally, in the event of malignancy confirmation through biopsy, the attention shifts to margin definition for surgical or treatment plan. Recognizing precise margins is crucial to reducing positive margin rates.

1.2 Computer-Aided Detection and Diagnosis

In this context, Artificial Intelligence (AI) emerges as a promising tool, capable of supporting the clinical pathway of patients with LC from diagnosis to

Table 1. Comparison with the state of the art.

Quality ass.: [5]	ResNet-18	22132 WL frames	F1-score = 80%	30 ms
Quality ass.: [6]	VGG16+clustering	720 NBI frames	F1-score = 95%	?
Detection: [13]	YOLOv5s	624 WL/NBI frames	AP = 63%	26 ms
Detection: [12]	YOLOv5s+m	4488 WL frames	AP = 74%	16 ms
Classification: [17]	DenseNet-121+attention	3057 WL frames	Accuracy = 73%	?
Classification: [21]	DenseNet-121	2254 WL frames	Accuracy = 86%	?
Segmentation: [24]	Multi-modal UNet	31543 WL/NBI frames	mIoU = 69%	26 ms

surgery. In a recent work [3], the role of AI and its potential impact in several settings of endoscopy for head and neck cancer was analyzed. Deep Learning (DL) approaches applied to endoscopic images have mainly been studied in relation to frame quality assessment for time savings during the review of endoscopic examinations and to reduce memory required for storage [4–7], lesion detection [8–13] and histology classification [14–21] to decrease the incidence of misdiagnosis and tumor segmentation to improve surgical outcomes [22–25]. In Table 1, we provided a general overview of the state-of-the-art works that most align with our purpose for each of the four addressed tasks. These approaches achieved encouraging results, but they still have shortcomings, including their focus on a restricted range of lesion categories and validation on either a small dataset or a single optical modality, i.e., standard White Light (WL) or NBI only. However, analyzing images of several disorders taken in both modes is relevant for transferring Computer-Aided Detection (CADe) and Diagnosis (CADx) approaches in clinical settings, where an otolaryngologist, during the endoscopic examination, might alternate between the two modalities by pressing a switch button. Moreover, the authors rarely consider the need to reach a reasonable trade-off between the complexity and robustness of the proposed DL methods and their actual implementation and integration in a clinically feasible and computationally efficient system.

Our work aims to advance the state of the art in this field by integrating multiple DL-based methods to handle the challenges experienced by clinicians over the course of LC clinical management, phase by phase. DL-based algorithms for informative frame selection, CADe, CADx and tumor segmentation were developed and embedded in a system mounted on a medical cart, compliant with the clinical practice thanks to real-time video processing capabilities and an intuitive graphical user interface for easy operation by non-technical users.

2 Material and Methods

2.1 Datasets

Considering the requirement for large and heterogeneous datasets to train deep neural networks, we collected sets of laryngeal endoscopic frames from 1,624

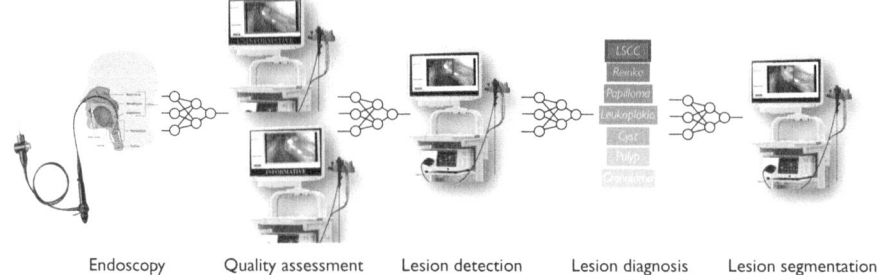

Fig. 1. Overview of the proposed system for less-experienced clinicians assistance in decision-making of larynx assessment. Starting from endoscopic videos, four tasks are addressed and automatized: image quality assessment, laryngeal lesion detection and classification for what concerns the outpatient examination, and segmentation of malignant tumors in subsequent clinical planning steps towards surgery.

patients enrolled in several clinical centers. The first center involved in this project was the Otorhinolaryngology Unit of the San Martino Hospital, University of Genoa, in Genoa, Italy[1]. Endoscopic videos were captured by clinical partners during the outpatient assessment of the laryngeal anatomical area using a high-definition video flexible naso pharynx laryngoscope[2] or during intraoperative procedures with rigid endoscopes coupled with high-definition camera[3]. This collection counted 12,193 White Light (WL) or NBI images with different resolutions from healthy subjects and patients with laryngeal lesions. Data included malignant lesions, i.e., Laryngeal Squamous Cell Carcinoma (LSCC), and benign lesions such as cysts, granuloma, leukoplakia, papilloma, polyps, and Reinke's edema. Another set used in this research was the Laryngoscope8 dataset, which is publicly available [17]. It comprised 3,057 WL laryngeal frames categorized into eight groups: glottic cancer, Reinke's edema, granuloma, vocal cord leukoplakia, vocal cord cysts, vocal cord nodules, and vocal cord polyps. The images were collected during head and neck surgery procedures at the Department of Otorhinolaryngology of the Sixth Medical Center of PLA General Hospital in Beijing, China. Similarly, the NBI-InfFrames publicly available dataset from [26] was considered. It comprises 720 frames from NBI videos of patients affected by LSCC. The images were captured with a resolution of 1920×1072 pixels, using a video flexible naso pharynx laryngoscope[2]. There are four types of frames in this collection: informative frames with proven good-quality information, dark frames due to underexposure, blurred frames, and those that exhibit either saliva

---

[1] This study was conducted following the principles of the Declaration of Helsinki and the local Institutional Review Board approval was obtained (CER Liguria: 169/2022).

[2] HD Video Rhino-laryngoscope Olympus ENF-VH, Olympus Medical System Corporation, Tokyo, Japan.

[3] 0°, 30°, or 70° rigid telescopes with HD camera head connected to a Visera Elite CLV-S190 light source, Olympus Medical System Corporation, Tokyo, Japan.

or specular reflections. The third clinical center enrolled was the ENT Center, Voice & Airway Clinic of the Sushrut Institute of Plastic Surgery & Super Specialty Hospital, Lucknow, India. Data contained 966 WL and NBI frames acquired with a high-definition video flexible naso pharynx laryngoscope[2] from both healthy subjects and patients who presented abnormalities. Moreover, 1,838 WL and NBI laryngeal images were collected at the Hospital Clínic of Barcelona, Barcelona, Spain, with the previously mentioned flexible laryngoscope[2]. Endoscopic frames, both informative and uninformative, belonged to patients with a healthy larynx or affected by several laryngeal conditions (LSCC, papilloma, polyps, Reinke's edema, granuloma, and nodule). Finally, the Unit of Otolaryngology of the Spedali Civili at the University of Brescia, Brescia, Italy, and the Unit of Otolaryngology of Severance Hospital, Yonsei University, Seoul, South Korea, provided their dataset from SCC cases. The former comprised 200 WL and NBI images, while the latter consisted of 156 WL and NBI images.

As diversified tasks require different annotation formats, the overall dataset was rearranged into four subsets related to tasks according to task-specific requirements: informativeness dataset, detection dataset, classification dataset, and segmentation dataset. Subsets were subsequently annotated by laryngologists with more than 5 years of experience in the field with appropriate labels for the task. Details about each subset are available in the following subsections. To reduce bias and prevent data leakage and overfitting, on-the-fly training-data augmentation, including image rotation, zoom, flipping, and shift, and train/test splitting strategy at patient level were applied.

2.2 Informativeness Network for Quality Assessment

The informativeness dataset was organized into 3126 frames for the training phase, 1317 frames for validation, and 704 frames for testing. For this task, selected WL and NBI data were labeled as "informative" if the image quality was deemed good, and "uninformative" whenever it was underexposed, blurred, affected by specular reflections, or presented occlusions due to saliva. ResNet-50 network was pre-trained with a pretext strategy based on $0°$, $90°$, $180°$, and $270°$ rotations of patches extracted from the laryngeal frames, and fine-tuned on the laryngeal dataset, resized to 512×512 pixels, for 200 epochs using the categorical cross-entropy loss, Adam optimizer, a batch size of 8, and an initial learning rate (lr_0) of 0.0001 as hyperparameters.

2.3 Lesion Detection Network

For laryngeal lesion detection, the dataset was prepared by asking the clinicians to annotate a bounding box around any suspicious lesion using the CVAT annotation software[4]. As the investigation of recent publications in endoscopic lesion detection resulted in an inclination toward the You Only Look Once (YOLO) models released by the Ultralytics team, we selected the YOLOv8nano variant,

---

[4] https://github.com/opencv/cvat (accessed February 28, 2023).

pre-trained on the ImageNet dataset, over its excellent trade-off between detection accuracy and inference time. Hence, annotations were obtained in the YOLO format (4 coordinates identifying the bounding box and class = "lesion"), and data were split into 5,128 for training, 770 for validation, and 337 test frames. After comprehensive hyperparameters tuning, we opted for the AdamW optimizer with a lr_0 equal to 0.001, a minimal learning rate of $0.1 \cdot lr_0$, image size of 640×640, autobatch and a value of 300 epoches for early stopping.

2.4 Lesion Classification Network

Two additional models were developed to classify laryngeal lesions into subclasses from endoscopic frames: the first model involved classifying three macro-classes (benign, malignant, and healthy), while the second aimed to classify frames into eight classes (cyst, granuloma, healthy, leukoplakia, papilloma, polyp, Reinke's edema, and LSCC). The classification dataset consists of 3,863 frames selected for training, 825 for validation, and 834 frames to test the developed networks. Considering the processing pipeline in Fig. 1, this task follows lesion detection. Therefore, all images were labeled with one of the eight classes above by expert otolaryngologists and cropped around the lesion according to the annotated bounding box. This functionality is primarily thought to provide decision support in outpatient examination, and extensive experimentation will assess the potential of this system to obviate the need for biopsy. ConvNextTiny pre-trained on ImageNet, which demonstrated superiority against other state-of-the-art models in a preliminary study, was fine-tuned on laryngeal data for 30 epochs, with an input image size of 256×256, batch size of 8, and Adam optimizer with a lr_0 equal to 0.001.

2.5 LSCC Segmentation Network

The final approach regards the implementation of laryngeal tumor margin identification for surgical excision. A DL model named SegMENT, developed [27], was considered for this task and was enhanced as in [25]. 3933 WL and NBI frames from LSCC cases were selected and cancer margins were carefully annotated via CVAT by our clinical collaborators. A split ratio of 90:10:10 was utilized to separate the training, validation and test frames. For the training of the enhanced network, the Dice loss function with the Adam optimizer, a lr_0 of 0.001, a batch size of 8 256×256-pixel images, and a number of epochs of 30 were used.

3 Results

3.1 Quality Assessment

The proposed model for quality assessment reached an F1-score of 96% on the test set (Table 2). Furthermore, the method allowed the informativeness of any frame to be predicted in less than 16 milliseconds (ms) on average with an

NVIDIA RTX A6000 GPU (48 GB of memory), demonstrating its potential for real-time applications. Figure 2 illustrates in the second column one example of a frame classified as uninformative due to blur and underexposure, and three examples of informative WL and NBI frames from healthy subjects and patients with benign or malignant disorders. The prediction is conveyed to the user thanks to the color of the edge box around the frame: red color corresponds to uninformative, and green to informative.

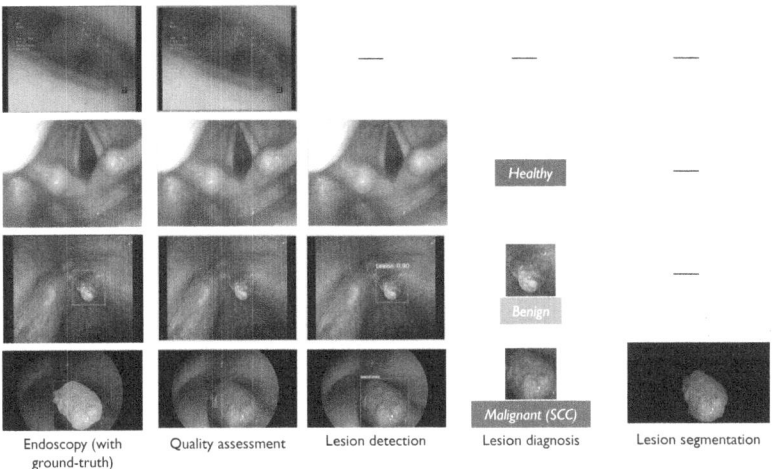

Fig. 2. Examples of task-related outcomes for four laryngeal frames. In the first line, the frame was correctly classified as uninformative (red edge box) due to blur and underexposure. In the other rows, frames were predicted as informative (green edge boxes) and the detection network was implemented to localize eventual lesions: the frame in the second row did not contain any lesions, so it was classified as healthy. The third frame, acquired with NBI modality in-office, represents an example of benign lesion, while for the last WL frame from intra-operative evaluation of a patient affected by LSCC, also the segmentation of the detected lesion was computed. (Color figure online)

3.2 Lesion Detection

YOLOv8nano achieved a mean Average Precision (AP) at Intersection over Union (IoU) = 0.50 equal to 73% when examined on test frames, maintaining the inference time below 20 ms with the same NVIDIA RTX A6000 GPU (Table 2). Some results of the detection task, exclusively applied to informative frames, are shown in the third column of Fig. 2: except for the case of healthy larynx in the second row, lesions were localized by a red bounding box superimposed on the NBI or WL frames.

3.3 Lesion Classification

The overall accuracy of the classification network for three classes was 88%, while for eight classes, the accuracy was 80% (Table 2). Concerning the three-class classification, occurrences for each one of the healthy, benign, and malignant classes, are reported in the fourth column of Fig. 2. The inference time during network testing on the GPU mentioned earlier was never greater than 10 ms.

3.4 LSCC Segmentation

The network proposed for segmenting cancerous lesions outperformed the other investigated DL models, obtaining 83% IoU, 83% Dice Similarity Coefficient (DSC), and an inference time always less than 40ms on the same GPU (Table 2). As illustrated in the last row of Fig. 2, when during inference a frame was identified as informative and if a malignant lesion was localized inside it, the segmentation mask was computed to define the precise boundaries of such tumor growth.

Table 2. For each task, network and dataset information, results, and speed of the selected DL-based approach are specified.

Task	Network	Dataset information	Results	Inference speed
Quality assessment	ResNet-50	5147 frames, labeled as informative or uninformative	F1-score = 96%, Precision = 95%, Recall = 97%	<16 ms
Lesion detection	YOLOv8n	6235 frames with bounding boxes annotated around each lesion	AP = 73%, Precision = 75%, Recall = 68%	<20 ms
Lesion classification	ConvNext-Tiny	5522 frames cropped based on the identified lesion and associated with one of the 3 or 8 classes	Accuracy $_{3-class}$ = 88%, $_{8-class}$ = 80%	<10 ms
LSCC segmentation	SegMENT-plus	3933 frames, for which clinicians have delineated the tumor margins	IoU = 83% DSC = 83%	<40 ms

4 Conclusions and Discussion

In the current clinical setting, the management of laryngeal diseases prioritizes the use of a multidisciplinary approach that involves flexible endoscopy before histopathological confirmation and any surgical intervention. Challenges can arise from difficulties in lesion detection and diagnosis, intensified by bad-quality data and clinician inexperience. AI has emerged as a promising tool for managing

LC, ranging from good-quality frame prediction to precise tumor segmentation for surgery. This work integrated and successfully tested several networks to provide a comprehensive decision support system for LC clinical workflow, both during outpatient examinations and surgery. After implementing four models for informative frame selection, lesion detection and classification, and LSCC segmentation, all of which revealed accurate results and inference times appropriate for real-time operation, we standardized the networks, developed with different Python DL libraries, by converting them into the ONNX format, and we tested them on external data, demonstrating the ability to generalize to previously unseen images. Then we installed the integrated system on a medical cart equipped with a powerful workstation, touchscreens, and a video grabber card for real-time acquisition of videos from any clinical endoscope. A novel intuitive GUI is being developed for operation, data acquisition, and annotations. Considering the stimulating results we observed in experiments conducted in our laboratory, clinical trials will commence soon and future work will exploit unsupervised approaches for domain adaptation and robustness improvement, and system optimization based on feedback from its real operational use and valuable new data.

Acknowledgments. Partially funded by the European Union - Horizon Europe project "AIRCARE - AI-augmented Robotics for CAncer point of caRE" (101137426), and by NextGenerationEU and the Ministry of University and Research (MUR), National Recovery and Resilience Plan (NRRP), Mission 4, Component 2, Investment 1.5, project "RAISE - Robotics and AI for Socio-economic Empowerment" (ECS00000035).

Disclosure of Interests. The authors have no competing interests to declare that are relevant to the content of this article.

References

1. Barsouk, A., Aluru, J.S., Rawla, P., Saginala, K., Barsouk, A.: Epidemiology, risk factors, and prevention of head and neck squamous cell carcinoma. Med. Sci. **11**(2) (2023)
2. Davaris, N., Voigt-Zimmermann, S., Kropf, S., Arens, C.: Flexible transnasal endoscopy with white light or narrow band imaging for the diagnosis of laryngeal malignancy: diagnostic value, observer variability and influence of previous laryngeal surgery. Eur. Arch. Otorhinolaryngol. **276**(2), 459–466 (2019)
3. Sampieri, C., et al.: Artificial intelligence for upper aerodigestive tract endoscopy and laryngoscopy: a guide for physicians and state-of-the-art review. Otolaryngol. Head Neck Surg. **169**(4), 811–829 (2023)
4. Patrini, I., Ruperti, M., Moccia, S., Mattos, L.S., Frontoni, E., De Momi, E.: Transfer learning for informative-frame selection in laryngoscopic videos through learned features. Med. Biol. Eng. Comput. **58**(6), 1225–1238 (2020)
5. Yao, P., et al.: Automatic classification of informative laryngoscopic images using deep learning. Laryngoscope Investig. Otolaryngol. **7**(2), 460–466 (2022)

6. Zhang, L., Wu, L., Wei, L., Wu, H., Lin, Y.: A novel framework of manifold learning cascade-clustering for the informative frame selection. Diagnostics **13**(6), 1151 (2023)
7. Baldini, C., et al.: An automated approach for real-time informative frames classification in laryngeal endoscopy using deep learning. Eur. Arch. Otorhinolaryngol. **281**, 4255–4264 (2024)
8. Cen, Q., Pan, Z., Li, Y., Ding, H.: Laryngeal tumor detection in endoscopic images based on convolutional neural network. In: 2019 IEEE 2nd International Conference on Electronic Information and Communication Technology (ICEICT). IEEE, Harbin (2019)
9. Luan, B., Sun, Y., Tong, C., Liu, Y., Liu, H.: R-FCN based laryngeal lesion detection. In: 12th International Symposium on Computational Intelligence and Design (ISCID), p. 2019. IEEE, Hangzhou (2019)
10. Yan, P., et al.: Automated detection of glottic laryngeal carcinoma in laryngoscopic images from a multicentre database using a convolutional neural network. Clin. Otolaryngol. **48**(3), 436–441 (2023)
11. Kim, G.H., Hwang, Y.J., Lee, H., Sung, E.S., Nam, K.W.: Convolutional neural network-based vocal cord tumor classification technique for home-based self-prescreening purpose. Biomed. Eng. Online **22**(1), 81 (2023)
12. Wellenstein, D.J., Woodburn, J., Marres, H.A.M., van den Broek, G.B.: Detection of laryngeal carcinoma during endoscopy using artificial intelligence. Head Neck **45**(9), 2217–2226 (2023)
13. Azam, M.A., et al.: Deep learning applied to white light and narrow band imaging videolaryngoscopy: toward real-time laryngeal cancer detection. Laryngoscope **132**(9), 1798–1806 (2022)
14. Xiong, H., et al.: Computer-aided diagnosis of laryngeal cancer via deep learning based on laryngoscopic images. EBioMedicine **48**, 92–99 (2019)
15. Dunham, M.E., Kong, K.A., McWhorter, A.J., Adkins, L.K.: Optical biopsy: automated classification of airway endoscopic findings using a convolutional neural network. Laryngoscope **132**(Suppl 4), S1–S8 (2022)
16. Ren, J., et al.: Automatic recognition of laryngoscopic images using a deep-learning technique. Laryngoscope **130**(11), E686–E693 (2020)
17. Yin, L., Liu, Y., Pei, M., Li, J., Wu, M., Jia, Y.: Laryngoscope8: laryngeal image dataset and classification of laryngeal disease based on attention mechanism. Pattern Recogn. Lett. **150**, 207–213 (2021)
18. Zhao, Q., et al.: Vocal cord lesions classification based on deep convolutional neural network and transfer learning. Med. Phys. **49**(1), 432–442 (2022)
19. Yao, P., et al.: A deep learning pipeline for automated classification of vocal fold polyps in flexible laryngoscopy. Eur. Arch. Otorhinolaryngol. **281**(4), 2055–2062 (2024)
20. You, Z., et al.: Vocal cord leukoplakia classification using deep learning models in white light and narrow band imaging endoscopy images. Head Neck **45**(12), 3129–3145 (2023)
21. Xu, Z.H., Fan, D.G., Huang, J.Q., Wang, J.W., Wang, Y., Li, Y.Z.: Computer-aided diagnosis of laryngeal cancer based on deep learning with laryngoscopic images. Diagnostics **13**(24), 3669 (2023)
22. Ji, B., et al.: A multi-scale recurrent fully convolution neural network for laryngeal leukoplakia segmentation. Biomed. Signal Process. Control **59**, 101913 (2020)
23. Zhou, L., et al.: Point-wise spatial network for identifying carcinoma at the upper digestive and respiratory tract. BMC Med. Imaging **23**(1), 140 (2023)

24. Li, Y., et al.: Real-time detection of laryngopharyngeal cancer using an artificial intelligence-assisted system with multimodal data. J. Transl. Med. **21**(1), 698 (2023)
25. Sampieri, C., et al.: Real-time laryngeal cancer boundaries delineation on white light and narrow-band imaging laryngoscopy with deep learning. Laryngoscope **134**(6), 2826–2834 (2024)
26. Moccia, S., et al.: Learning-based classification of informative laryngoscopic frames. Comput. Methods Programs Biomed. **158**, 21–30 (2018)
27. Azam, M.A., et al.: Videomics of the upper aero-digestive tract cancer: deep learning applied to white light and narrow band imaging for automatic segmentation of endoscopic images. Front. Oncol. **12**, 900451 (2022)

UltraWeak: Enhancing Breast Ultrasound Cancer Detection with Deformable DETR and Weak Supervision

Ufaq Khan[1(✉)], Umair Nawaz[1], and Abdulmotaleb E. Saddik[1,2]

[1] Mohamed Bin Zayed University of Artificial Intelligence,
Abu Dhabi, United Arab Emirates
ufaq.khan@mbzuai.ac.ae
[2] University of Ottawa, Ottawa, Canada

Abstract. In breast ultrasound imaging, the scarcity of detailed annotations poses a major barrier to developing robust object detection models. This challenge is compounded by the high intra-class variability, where different slices of a 3D object can appear drastically different in 2D images, and low inter-class variance, where pathological features are often small and subtle compared to the rest of the image. These factors make it difficult to train models that require precise bounding box annotations or extensive labeled datasets. Addressing these issues, this study introduces a novel weakly supervised object detection (WSOD) model that capitalizes on image-level labels, which are more readily available and require significantly less effort from medical professionals. Our approach integrates Multi-Instance Learning (MIL) and Self-Supervised Learning (SSL) within a Deformable DETR framework, aiming to focus the model's attention on relevant regions without detailed annotations. Tested on the two publicly available datasets, our model demonstrates significant improvements in mean average precision (mAP) and recall, surpassing existing state-of-the-art methods. Ablation studies confirm the essential importance of MIL and SSL in enhancing detection accuracy, validating our model as a potent solution for overcoming data scarcity in medical imaging.

Keywords: Weakly Supervised Object Detection · Self-Supervised Learning · Breast Cancer Detection · Ultrasound Imaging

1 Introduction

In medical imaging, the detection and accurate localization of anatomical structures and pathological findings are essential for diagnosis, treatment planning, and patient management [20]. However, the application of conventional supervised deep learning techniques [6,11] is often hampered by the high cost and labor-intensive nature of obtaining precise annotations. Such detailed annotations require significant expert time and are prone to variability depending on the annotator's expertise. Weakly supervised object detection (WSOD) [19] emerges as a compelling solution by reducing the dependency on detailed annotations and leveraging weaker forms, such as image-level labels, partial annotations, or noisy labeled data.

Despite its potential, WSOD in medical imaging faces unique challenges [14]. First, medical images often exhibit high intra-class and low inter-class variability, complicating distinguishing between categories or structures based solely on weak labels. Furthermore, the critical nature of medical diagnosis demands high accuracy and reliability from detection models [7], which are traditionally challenging to achieve under weak supervision.

Recent advancements in machine learning have begun addressing these challenges. For instance, innovations in transfer learning have shown promise in adapting models trained on richly annotated non-medical datasets to medical applications [15]. Techniques such as multi-instance learning (MIL) [9,12] have also been adapted to leverage ambiguous or imprecise labels, particularly in scenarios where only some regions of interest are annotated within an image. Moreover, semi-supervised and self-supervised learning paradigms have been explored to utilize unlabeled data, enhancing the training process under limited annotation conditions [5,10]. Tang et al. [17] enhance weakly supervised object detection with their Multiple Instance Detection Network, which dynamically refines instance classifiers during training. This refinement incorporates mechanisms that could be seen as an early form of self-supervised learning. Our approach to self-supervised learning was inspired by this innovative method.

This study introduces an innovative approach to Weakly Supervised Object Detection (WSOD) in medical imaging by integrating the Deformable Detection Transformer (DETR) [22] with Multiple Instance Learning (MIL) [9] and self-supervised learning technique [10] (Fig. 1). This method is specifically designed to enhance the detection and localization of pathological features such as breast cancer lesions in ultrasound images. Deformable DETR, which incorporates deformable attention mechanisms, dynamically focuses on relevant image features, improving efficiency and accuracy in identifying regions of interest critical for medical diagnosis. MIL handles the ambiguities of weakly labeled data by treating each image as a bag of instances, allowing the model to learn from region proposals without precise object annotations. This is particularly beneficial in medical settings where detailed annotations are costly and challenging to obtain.

To further refine the model's diagnostic capabilities, self-supervised learning techniques are utilized [17]. This learning phase focuses on refining instance representations based on the model's predictions, creating a feedback loop where the model iteratively improves its predictive accuracy. Unlike typical self-supervised tasks like rotation prediction or image reconstruction, this method leverages the inherent predictions and errors of the model to guide its continuous learning process, enhancing its ability to differentiate subtle pathological changes within the images. The key contributions of this paper are:

- We introduce a novel adaptation of Deformable DETR integrated with Multiple Instance Learning (MIL), specifically tailored for breast ultrasound

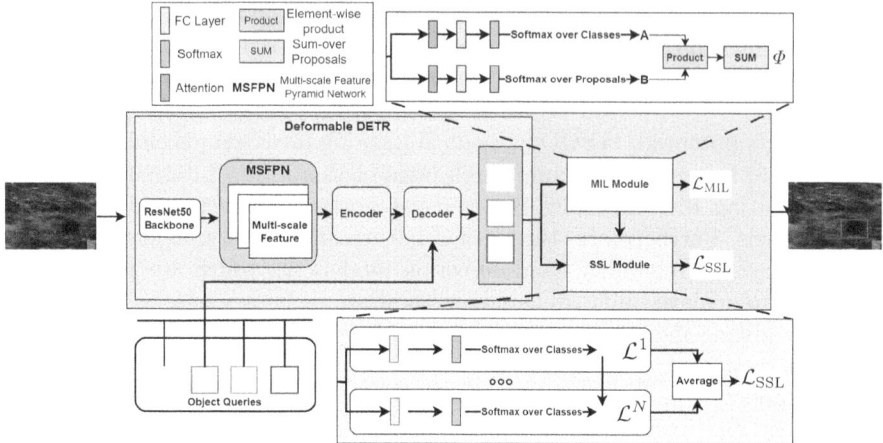

Fig. 1. The workflow for Weakly Supervised Object Detection (WSOD) in breast cancer detection begins with inputting raw breast ultrasound images into the system. These images undergo feature extraction via a Deformable DETR model to identify possible regions of interest (bounding boxes). Using Multiple Instance Learning (MIL), these proposals are classified within bags as malignant or benign. A self-supervised instance learning module refines the classification and detection process, leading to the final detection and classification of breast lesions.

imagery. This approach leverages deformable attention to enhance detection accuracy significantly under weakly supervised conditions.
- Our research advances self-supervised learning strategies to refine detection, demonstrating marked improvements in model robustness and predictive accuracy over traditional WSOD methods.
- We validate our model's effectiveness through rigorous testing on two distinct datasets for breast cancer detection in ultrasound images, confirming its potential for real-world medical diagnostics.

2 Methodology

In this work, we propose a comprehensive framework combining Deformable DETR with MIL and self-supervised learning to address the challenges of weakly supervised object detection in medical imaging, particularly focusing on detecting and localizing breast cancer lesions in ultrasound images.

Deformable DETR: Deformable DETR [22] modifies the standard DETR [4] by introducing deformable attention mechanisms to make it more efficient. These mechanisms allow the model to focus dynamically on a limited number of key sampling points rather than the entire image, which enhances the model's efficiency and effectiveness in identifying regions of interest. We further introduce an adaptive proposal generation mechanism within the Deformable DETR framework. This mechanism adjusts the density and scale of the region proposals based on the detected feature complexity within specific image areas. This approach is introduced to suit the heterogeneous nature of medical images.

Let $\mathbf{X}$ be the input feature map from the backbone network, and $\mathbf{z}$ be the set of learnable parameters defining the deformable attention. The deformable attention operation in Eq. 1 can be formulated as:

$$\mathbf{Y} = \text{DeformAttn}(\mathbf{X}, \mathbf{z}) = \sum_{k=1}^{K} \mathbf{W}_k \cdot \mathbf{X}(\phi_k(\mathbf{z})) \tag{1}$$

where ϕ_k represents the adaptive sampling function for the k-th key point, and $\mathbf{W}_k$ are the weights associated with each key point.

Multiple Instance Learning (MIL): In our adapted framework, the decoder of the modified Deformable DETR outputs a set of feature vectors, where each vector corresponds to a potential region of interest proposed by the model. These vectors are processed by distinct, attention-focused, and fully connected layers to derive classification scores for each category and their respective detection scores. Let the feature vectors be denoted as $\mathbf{F} \in \mathbb{R}^{R \times d}$, where R represents the number of proposals generated by the detector, and d is the dimensionality of each vector. Each vector $\mathbf{F}_r$ undergoes transformation through fully connected layers to produce two matrices: $\mathbf{A} \in \mathbb{R}^{R \times N_c}$ for class scores, and $\mathbf{B} \in \mathbb{R}^{R \times N_c}$ for detection scores, with N_c being the number of target classes.

The class-specific probabilities $\mathbf{P}$ and detection scores $\mathbf{Q}$ are computed via softmax operations in Eq. 2:

$$\mathbf{P}_{ij} = \frac{\exp(\mathbf{A}_{ij})}{\sum_{k=1}^{N_c} \exp(\mathbf{A}_{ik})}, \quad \mathbf{Q}_{ij} = \frac{\exp(\mathbf{B}_{ij})}{\sum_{k=1}^{N_c} \exp(\mathbf{B}_{ik})} \tag{2}$$

where $\mathbf{P}_{ij}$ and $\mathbf{Q}_{ij}$ represent the probability of the i-th proposal belonging to the j-th category and the detection confidence score for the j-th category in the i-th proposal, respectively. The overall image classification score Φ_j for each category j is derived by combining the class-specific probabilities and detection scores across all proposals: $\Phi_j = \sum_{i=1}^{R} \mathbf{P}_{ij} \cdot \mathbf{Q}_{ij}$. The MIL loss is computed using a binary cross-entropy loss between the predicted image-level labels and the actual image labels, which is given in Eq. 3:

$$\mathcal{L}_{MIL} = -\sum_{j=1}^{N_c} [y_j \log(\Phi_j) + (1 - y_j) \log(1 - \Phi_j)] \tag{3}$$

where y_j is the ground truth label for the j-th category at the image level, indicating the presence or absence of the corresponding disease.

Self-supervised Learning Instance: Building upon the methodologies discussed in [8,17], our approach devises a self-supervised instance learning module comprising several blocks to refine instance scores using instance-level feedback. This feedback is derived from the outputs of prior processing stages, enabling a more precise and adaptable learning framework. Our model features a series of N_r blocks within the self-supervised learning framework, each equipped with a fully connected (FC) layer. These blocks iteratively refine the instance scores based on feedback from the preceding layer, accurately enhancing the model's ability to discern between pathological and non-pathological regions.

Each block computes instance scores using a class-wise softmax function, resulting in a scoring matrix $\mathbf{x}_n \in \mathbb{R}^{R \times (N_c+1)}$ at the n-th block, where $N_c + 1$ accounts for the additional background class. The supervision for each block n derives from the scores generated by the previous block $\mathbf{x}_{(n-1)}$. Specifically, the initial supervision for the first block is provided by the MIL head, forming a foundational layer of instance predictions: $\mathbf{m}_j^n = \arg\max_i \mathbf{x}_{ij}^{n-1}$.

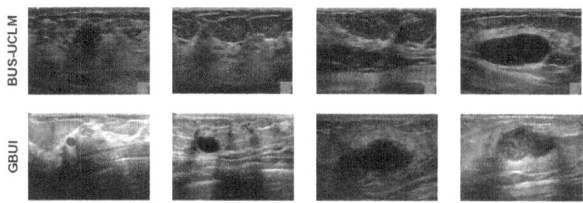

Fig. 2. Sample images for each of the dataset

The pseudo-labels $\hat{\mathbf{y}}_n$ for the instances are determined based on the overlap criterion with the highest-scoring instance from the preceding layer's outputs, employing the Intersection over Union (IoU) as the metric in Eq. 4:

$$\hat{y}_{ij}^n = \begin{cases} 1 & \text{if } \text{IoU}(p_j, p_{m_j^n}) \geq \tau \\ 0 & \text{otherwise} \end{cases} \quad (4)$$

where τ is a predefined threshold for determining positive instances.

The loss for this self-supervised learning component is calculated over all blocks and instances, aimed at minimizing the difference between the predicted instance scores and the derived pseudo-labels in Eq. 5:

$$\mathcal{L}_{\text{SSL}} = -\frac{1}{N_r} \sum_{n=1}^{N_r} \frac{1}{R} \sum_{i=1}^{R} \sum_{j=1}^{N_c+1} w_i^n \hat{y}_{ij}^n \log x_{ij}^n \quad (5)$$

Here, x_{ij}^n denotes the score of the i-th instance for the j-th class at layer n, and the loss weight $w_{n,i}$ is defined based on the score of the most significant instance from the previous layer, stabilizing the learning process: $w_i^n = x_{i,m_j^n}^{(n-1)}$. The overall loss function integrates the MIL loss with the instance refinement loss, balanced by a scaling factor λ to accommodate the different magnitudes of contributions from each component: $\mathcal{L}_{total} = \mathcal{L}_{\text{MIL}} + \lambda \mathcal{L}_{\text{SSL}}$.

This formulation ensures that the learning process is sensitive to the weakly labeled overarching image-level information and the refined instance-level details, fostering a comprehensive understanding critical for effective diagnosis in medical imaging.

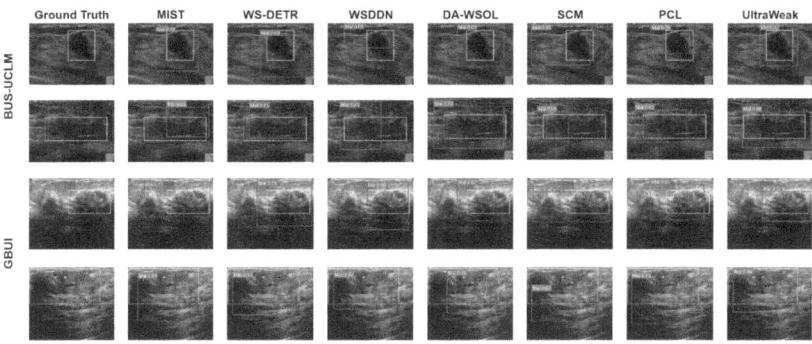

Fig. 3. Visual representation of the WSOD results obtained from various models on both the datasets. Ground truth bounding boxes are shown in green, whereas the model predictions are shown in red. (Color figure online)

3 Experimental Setup

Datasets: Our research utilizes two distinct datasets of breast ultrasound images, each chosen for its relevance to different aspects of breast condition analysis (Fig. 2). The first dataset, the BUS-UCLM dataset [18], is designed specifically for segmenting breast ultrasound lesions. It includes 683 images from 38 patients, providing a range of breast conditions. The dataset features RGB segmentation masks, with benign lesions marked in green and malignant lesions in red. Additionally, we also used another breast ultrasound dataset [1], which we named Generic Breast Ultrasound Image (GBUI), encompassing 780 images from 600 patients aged between 25 and 75 years. This extensive collection categorizes images into normal, benign, and malignant classes. The images are standardized at a resolution of 500 × 500 pixels. The bounding boxes were extracted from these mask images for our evaluation purposes for both datasets. Since both datasets encompass images from different patients, we use the 5-fold cross-validation approach at the patient level to avoid any data leakage. In this way, the images of a patient were split either in the train or validation set.

Preprocessing Steps: All images were resized to a uniform dimension of 224 × 224 pixels to standardize input size. Intensity normalization was applied to each image to enhance contrast and reduce variability in lighting and exposure. Additionally, data augmentation techniques such as random rotations, flipping, and scaling were employed to increase the robustness of the model against variations in image acquisition angles and scales.

Experimental Configurations: The Deformable DETR model was trained with a ResNet-50 backbone pre-trained on the ImageNet dataset to leverage transfer learning for initial feature extraction. The model includes N_r self supervised blocks, each with a fully connected layer designed to refine the detection and classification of pathological features. The number of object queries was set to 10, considering our dataset's typical number of regions of interest per image.

Table 1. Performance Comparison of WSOD Methods on Breast Ultrasound datasets using mean average precision ($mAP_{0.25}$)

Method	BUS-UCLM	GBUI
MIST [13]	0.465 ± 0.080	0.430 ± 0.058
WS-DETR [8]	0.489 ± 0.045	0.455 ± 0.067
WSDDN [3]	0.425 ± 0.065	0.398 ± 0.049
DA-WSOL [21]	0.492 ± 0.055	0.448 ± 0.060
SCM [2]	0.032 ± 0.014	0.025 ± 0.009
PCL [16]	0.448 ± 0.072	0.410 ± 0.054
UltraWeak (Ours)	0.535 ± 0.040	0.498 ± 0.052

The model was trained using a batch size of 16 for 30 epochs using the Adam optimizer. The initial learning rate was set at 0.0001, with a decay factor of 0.1 applied every 4 epochs to gradually reduce the learning rate and stabilize training. The combined loss function $L = L_{\text{MIL}} + \lambda L_{\text{ins}}$ was used, where L_{MIL} is the multiple instance learning loss and L_{ins} is the self-supervised instance refinement loss. The scaling factor λ was empirically set to 0.5 to balance the contributions from the MIL component and the instance refinement component effectively. All the experiments were conducted using a GPU-powered cluster with an NVIDIA Tesla A100 GPU and 128 GB of RAM.

Table 2. Comparison of UltraWeak to other baselines on BUS-UCLM and GBUI datasets for classification task

Method	BUS-UCLM			GBUI		
	Accuracy	Precision	Recall	Accuracy	Precision	Recall
MIST [13]	0.525 ± 0.030	0.510 ± 0.028	0.760 ± 0.045	0.515 ± 0.032	0.500 ± 0.031	0.750 ± 0.048
WS-DETR [8]	0.650 ± 0.035	0.635 ± 0.033	0.845 ± 0.038	0.640 ± 0.036	0.625 ± 0.035	0.835 ± 0.040
WSDDN [3]	0.575 ± 0.034	0.560 ± 0.032	0.805 ± 0.046	0.565 ± 0.035	0.550 ± 0.034	0.795 ± 0.045
DA-WSOL [21]	0.685 ± 0.027	0.670 ± 0.025	0.880 ± 0.032	0.675 ± 0.028	0.660 ± 0.027	0.870 ± 0.033
SCM [2]	0.455 ± 0.040	0.440 ± 0.042	0.695 ± 0.060	0.445 ± 0.041	0.430 ± 0.043	0.685 ± 0.058
PCL [16]	0.600 ± 0.038	0.585 ± 0.037	0.780 ± 0.050	0.590 ± 0.039	0.575 ± 0.038	0.770 ± 0.052
UltraWeak	0.720 ± 0.025	0.705 ± 0.023	0.915 ± 0.020	0.700 ± 0.026	0.685 ± 0.024	0.905 ± 0.022

4 Experimental Results

This section presents a quantitative and qualitative evaluation of our proposed weakly supervised object detection (WSOD) method, named UltraWeak, benchmarked against current state-of-the-art techniques. We assess the performance using the mean average precision (mAP) across two distinct datasets, as shown in Table 1. For WSOD, our method achieved the highest mAP score of 0.535 ± 0.040 on the BUS-UCLM dataset, demonstrating superior lesion detection using image labels only. Notable comparisons include WS-DETR with an mAP of 0.489 ± 0.045 and the SCM method, which significantly underperformed at 0.032 ± 0.014. Similarly, our method also outperformed others on the GBUI dataset with an mAP of 0.498 ± 0.052. WS-DETR showed competitive performance with an mAP of 0.455 ± 0.067, while SCM remained less effective with 0.025 ± 0.009.

Besides object detection, our model demonstrates robust performance in image classification tasks. Leveraging weak labels, the classification module was evaluated on the same datasets as shown in Table 2, focusing on metrics such as accuracy, precision, and recall to provide a comprehensive understanding of its effectiveness. For the BUS-UCLM dataset, the model achieved an accuracy of 0.720, precision of 0.705, and recall of 0.915, showcasing its ability to identify lesions with high reliability accurately. Similarly, on the GBUI dataset, the accuracy was 0.700, with a precision of 0.685 and a recall of 0.905, confirming the model's consistent performance across different imaging contexts.

In our qualitative analysis, Fig. 3 presents visual comparisons of cancer detection across different models. These results were obtained using a confidence score threshold of 50%. Conventional models like SCM, WS-DETR, and WSDDN often struggle with generalization, producing imprecise or overly loose bounding boxes. In contrast, our UltraWeak model excels, consistently generating well-fitted bounding boxes around target lesions with higher confidence scores. This precision in localization is crucial in clinical settings, where accurate detection directly influences the diagnostic utility and effectiveness of the imaging process. The results underscore the advanced capability of UltraWeak to handle real-world diagnostic challenges, offering significant improvements over existing solutions. Moreover, the key advantages of UltraWeak include reduced annotation effort, scalability to large datasets, and improved diagnostic accuracy. In summary, the UltraWeak approach offers a promising solution for breast ultrasound diagnostics with less reliance on detailed annotations, and future work should focus on real-world clinical trials to validate its effectiveness.

5 Ablation Study

Our ablation study assesses the importance of individual components within our weakly supervised object detection model. Results in Table 3 show significant declines in mAP and recall when the MIL head and SSL components are removed, indicating their critical contributions to model performance. Changes in the backbone network and optimizer had less impact, underscoring the robustness of the core model architecture. These insights guide effective enhancements for medical imaging applications.

Table 3. Ablation study results on BUS-UCLM and GBUI Datasets

Model Configuration	BUS-UCLM		GBUI	
	mAP	Recall	mAP	Recall
FullModel	0.535 ± 0.040	0.915 ± 0.030	0.498 ± 0.052	0.905 ± 0.035
Without MIL	0.498 ± 0.045	0.875 ± 0.035	0.462 ± 0.050	0.865 ± 0.040
Without SSL	0.510 ± 0.042	0.900 ± 0.025	0.475 ± 0.048	0.885 ± 0.030
Backbone (ResNet18)	0.520 ± 0.042	0.900 ± 0.035	0.485 ± 0.047	0.890 ± 0.040
Optimizer (SGD)	0.525 ± 0.038	0.905 ± 0.032	0.490 ± 0.045	0.895 ± 0.038

6 Conclusion

This study introduces an advanced weakly supervised object detection model that demonstrates superior performance in lesion detection within breast ultrasound images. Key findings from evaluations on the BUS-UCLM and GBUI datasets highlight our method's enhanced performance, which is crucial for ultrasound diagnostic accuracy. Ablation studies confirm the importance of MIL and SSL components in achieving these results. Future work will aim to expand these techniques to broader medical imaging applications, reinforcing the potential of WSOD models in improving diagnostic procedures.

Disclosure of Interests. The authors have no competing interests to declare that are relevant to the content of this article.

References

1. Al-Dhabyani, W., Gomaa, M., Khaled, H., Fahmy, A.: Dataset of breast ultrasound images. Data Brief **28**, 104863 (2020)
2. Bai, H., Zhang, R., Wang, J., Wan, X.: Weakly supervised object localization via transformer with implicit spatial calibration. In: Avidan, S., Brostow, G., Cissé, M., Farinella, G.M., Hassner, T. (eds.) ECCV 2022. LNCS, vol. 13669, pp. 612–628. Springer, Cham (2022). https://doi.org/10.1007/978-3-031-20077-9_36
3. Bilen, H., Vedaldi, A.: Weakly supervised deep detection networks. In: Proceedings of the IEEE Conference on Computer Vision and Pattern Recognition, pp. 2846–2854 (2016)
4. Carion, N., Massa, F., Synnaeve, G., Usunier, N., Kirillov, A., Zagoruyko, S.: End-to-end object detection with transformers. In: Vedaldi, A., Bischof, H., Brox, T., Frahm, J.-M. (eds.) ECCV 2020. LNCS, vol. 12346, pp. 213–229. Springer, Cham (2020). https://doi.org/10.1007/978-3-030-58452-8_13
5. Jiao, R.: Learning with limited annotations: a survey on deep semi-supervised learning for medical image segmentation. Comput. Biol. Med., 107840 (2023)
6. Kim, J., et al.: Weakly-supervised deep learning for ultrasound diagnosis of breast cancer. Sci. Rep. **11**(1), 24382 (2021)
7. Kumar, Y., Koul, A., Singla, R., Ijaz, M.F.: Artificial intelligence in disease diagnosis: a systematic literature review, synthesizing framework and future research agenda. J. Ambient. Intell. Humaniz. Comput. **14**(7), 8459–8486 (2023)

8. LaBonte, T., Song, Y., Wang, X., Vineet, V., Joshi, N.: Scaling novel object detection with weakly supervised detection transformers. In: Proceedings of the IEEE/CVF Winter Conference on Applications of Computer Vision, pp. 85–96 (2023)
9. Lerousseau, M., et al.: Weakly supervised multiple instance learning histopathological tumor segmentation. In: Martel, A.L., et al. (eds.) MICCAI 2020. LNCS, vol. 12265, pp. 470–479. Springer, Cham (2020). https://doi.org/10.1007/978-3-030-59722-1_45
10. Ouyang, C., Biffi, C., Chen, C., Kart, T., Qiu, H., Rueckert, D.: Self-supervised learning for few-shot medical image segmentation. IEEE Trans. Med. Imaging **41**(7), 1837–1848 (2022)
11. Peng, J., Wang, Y.: Medical image segmentation with limited supervision: a review of deep network models. IEEE Access **9**, 36827–36851 (2021)
12. Qian, Z., et al.: Transformer based multiple instance learning for weakly supervised histopathology image segmentation. In: Wang, L., Dou, Q., Fletcher, P.T., Speidel, S., Li, S. (eds.) MICCAI 2022. LNCS, vol. 13432, pp. 160–170. Springer, Cham (2022). https://doi.org/10.1007/978-3-031-16434-7_16
13. Ren, Z., et al.: Instance-aware, context-focused, and memory-efficient weakly supervised object detection. In: Proceedings of the IEEE/CVF Conference on Computer Vision and Pattern Recognition, pp. 10598–10607 (2020)
14. Shao, F., et al.: Deep learning for weakly-supervised object detection and localization: a survey. Neurocomputing **496**, 192–207 (2022)
15. Shin, H.-C., et al.: Deep convolutional neural networks for computer-aided detection: CNN architectures, dataset characteristics and transfer learning. IEEE Trans. Med. Imaging **35**(5), 1285–1298 (2016)
16. Tang, P., et al.: PCL: proposal cluster learning for weakly supervised object detection. IEEE Trans. Pattern Anal. Mach. Intell. **42**(1), 176–191 (2018)
17. Tang, P., Wang, X., Bai, X., Liu, W.: Multiple instance detection network with online instance classifier refinement. In: Proceedings of the IEEE Conference on Computer Vision and Pattern Recognition, pp. 2843–2851 (2017)
18. Vallez, N., Bueno, G., Deniz, O., Rienda, M.A., Pastor, C.: BUS-UCLM: breast ultrasound lesion segmentation dataset (2024). https://data.mendeley.com/datasets/7fvgj4jsp7/1
19. Zhang, D., Han, J., Cheng, G., Yang, M.-H.: Weakly supervised object localization and detection: a survey. IEEE Trans. Pattern Anal. Mach. Intell. **44**(9), 5866–5885 (2022)
20. Zhao, Y., et al.: Deep learning solution for medical image localization and orientation detection. Med. Image Anal. **81**, 102529 (2022)
21. Zhu, L., She, Q., Chen, Q., You, Y., Wang, B., Lu, Y.: Weakly supervised object localization as domain adaptation. In: Proceedings of the IEEE/CVF Conference on Computer Vision and Pattern Recognition, pp. 14637–14646 (2022)
22. Zhu, X., Su, W., Lu, L., Li, B., Wang, X., Dai, J.: Deformable DETR: deformable transformers for end-to-end object detection. arXiv preprint arXiv:2010.04159 (2020)

SelectiveKD: A Semi-supervised Framework for Cancer Detection in DBT Through Knowledge Distillation and Pseudo-labeling

Laurent Dillard, Hyeonsoo Lee, Weonsuk Lee, Tae Soo Kim, Ali Diba[✉], and Thijs Kooi

Lunit Inc, Seoul, Republic of Korea
{laurent.dillard,hslee,iwonseok5762,taesoo.kim,ali,tkooi}@lunit.io

Abstract. When developing Computer Aided Detection (CAD) systems for Digital Breast Tomosynthesis (DBT), the complexity arising from the volumetric nature of the modality poses significant technical challenges for obtaining large-scale accurate annotations. Without access to large-scale annotations, the resulting model may not generalize to different domains. Given the costly nature of obtaining DBT annotations, how to effectively increase the amount of data used for training DBT CAD systems remains an open challenge.

In this paper, we present SelectiveKD, a semi-supervised learning framework for building cancer detection models for DBT, which only requires a limited number of annotated slices to reach high performance. We achieve this by utilizing unlabeled slices available in a DBT stack through a knowledge distillation framework in which the teacher model provides a supervisory signal to the student model for all slices in the DBT volume. Our framework mitigates the potential noise in the supervisory signal from a sub-optimal teacher by implementing a selective dataset expansion strategy using pseudo labels.

We evaluate our approach with a large-scale real-world dataset of over 10,000 DBT exams collected from multiple device manufacturers and locations. The resulting SelectiveKD process effectively utilizes unannotated slices from a DBT stack, leading to significantly improved cancer classification performance (AUC) and generalization performance.

Keywords: Computer-aided diagnosis · machine learning · Semi-supervised Learning · Annotation efficiency

---

L. Dillard, H. Lee and W. Lee—These authors contributed equally to this work.

Supplementary Information The online version contains supplementary material available at https://doi.org/10.1007/978-3-031-73376-5_15.

1 Introduction

Digital Breast Tomosynthesis (DBT) offers three-dimensional imaging for breast cancer screening which improves over the traditional 2D mammography by better detecting cancer especially from dense breasts [1,8,12,16], lowering false-positive recall rates [2,17] and demonstrating overall superior cancer detection rates [6]. Because images are taken from multiple angles, lesions visibility is enhanced. However, the volumetric nature of the data can also increase the radiologists' workload caused by longer reading time as they have to go through an entire stack of images [19]. Recent advances in deep learning-based Computer-Aided Detection (CAD) systems for DBT offer a promising solution to mitigate this.

Deep learning-based CAD systems for breast cancer detection in DBT have been an active area of research [9,11,14,18,22]. However, existing CAD approaches for DBT also face challenges from the complexity arising from the volumetric nature of the modality. Each DBT stack comprises roughly 40 - 80 slices for each view, resulting in hundreds of images for each exam. Moreover, slices of interest (i.e., with a suspicious lesion) comprise only a small portion of a given DBT stack. As a result, collecting large-scale accurate annotations for marking such regions of interest poses a significant technical challenge. As a compromise, practitioners often build DBT CAD systems by annotating only a sparse set of slices of interest by marking them as positive and treating all other slices as negative [21,24]. This practice limits the scale of annotated datasets for DBT and introduces potential noise in the annotations. Without accurate large-scale annotations, deep learning models over-fit easily, ultimately hindering their clinical value as CAD [20].

In this paper, we present a semi-supervised learning framework, *SelectiveKD*, for building a cancer detection model that leverages unannotated slices in a DBT stack. Inspired by recent advancements in semi-supervised learning [4,5,11,18,23], we build upon the concept of Knowledge Distillation (KD) [10] to be able to use all the unlabeled slices from a DBT exam. We first obtain a teacher model using available annotated slices, which provide a supervisory signal for the student model, which trains using both annotated and unannotated slices.

In doing so, some noise can be introduced into the training process due to the teacher's inherent sub-optimal performance. Our framework mitigates this using Pseudo Labels (PL) generated with the teacher [3,13] to select which unannotated slices to include in training. We empirically show that PLs correct the noise introduced in the KD process. Our framework effectively combines KD and PL to leverage all slices available in a DBT volume using only a limited number of annotated slices, leading to significantly improved cancer detection performance.

We validate our framework using a large real-world dataset of over 10,000 DBT exams, collected from multiple device manufacturers (i.e., Hologic, GE, Siemens) and institutions. We demonstrate that the presented approach not only improves detection performance but also the model's ability to generalize across different manufacturers. We show that our framework achieves this generalization capacity *without* requiring annotations from the target device manufacturers. Further, by effectively using unlabeled images, we can actually reach the same level of cancer detection performance with considerably fewer labeled images, leading to large potential annotation cost savings for building practical CAD systems for DBT (Fig. 1).

2 Method

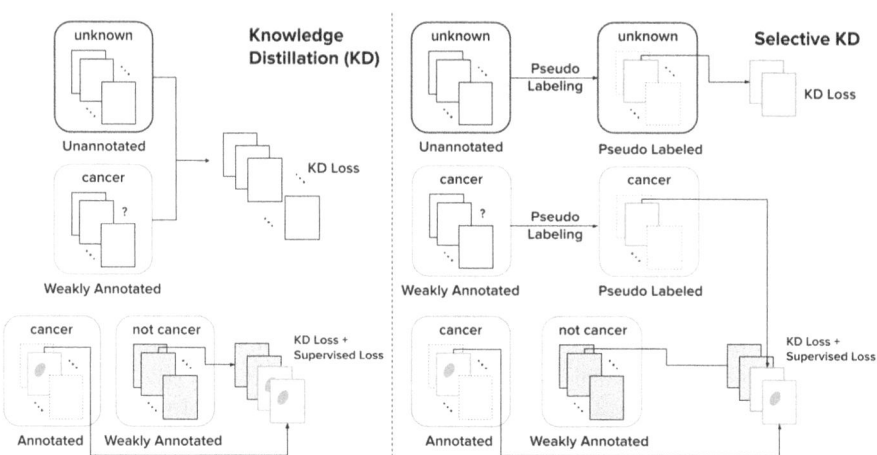

Fig. 1. Overview of the SelectiveKD framework: Illustrating the application of knowledge distillation (KD) and SelectiveKD across varying levels of annotation of DBT slices. SelectiveKD leverages pseudo-labeling to select a few slices for training, minimizing the potential noise introduced by standard KD and reinforcing weak annotations. The approach allows use of all slices in the DBT stack during training.

2.1 Practical Annotation of DBT Data

A Digital breast tomosynthesis (DBT) exam contains multiple stacks of high resolution images reconstructed from X-ray images taken from different angles. The task of annotating every single image within a DBT stack, typically consisting of 40 to 80 slices, is prohibitively time-consuming and expensive. Consequently, alternative annotation strategies are used.

One strategy involves selectively annotating the slice where the lesion is most clearly visible, marking it with a contour annotation. This method reduces the annotation workload, however, a thorough examination of each slice in the DBT stack is still required to identify the most representative one for annotation.

Although a single slice is annotated per DBT stack, it is safe to consider the lesion is roughly in the same location in adjacent slices; therefore, each stack can yield up to 3 annotated slices. In this work, annotated cases refer to cases annotated using this strategy. For annotated DBT stacks without lesions, all slices can be used for training. Still, in order to avoid a large imbalance between positive and negative slices, those stacks are sub-sampled using strided sampling.

An alternative approach is to collect weak annotations by utilizing other diagnostic results, such as ultrasound or biopsy. This approach can only yield breast-level annotation indicating the presence or absence of cancer. These weak annotations are difficult to leverage for model training due to the variable visibility of lesions across the stack of slices. Even in DBT stacks confirmed to exhibit cancerous lesions, pinpointing the specific slices where the lesions are detectable remains elusive. Note, however, that for DBT stacks confirmed to be cancer-negative, we can safely consider all slices to be negative and use them in training the same way as for fully annotated stacks.

2.2 Incorporating Pseudo-Labeling and Knowledge Distillation

To address the challenge of limited annotations in DBT data for classification and segmentation tasks, our methodology leverages unannotated data through pseudo-labeling (PL) and knowledge distillation (KD), carefully navigating the inherent risk of introducing noise from automatic labeling.

Refined Learning Through Knowledge Distillation: Confronting sparse annotations, our approach initially considers KD as a primary method. The model's training is refined through KD by leveraging the probabilistic outputs (soft labels) produced by the teacher model, which is pre-trained on a small annotated dataset. Unlike hard labels, soft labels provide a detailed continuum of information. This approach enriches the student model's training by incorporating the teacher's knowledge rather than just binary outcomes.

Selective Dataset Expansion through Pseudo-Labeling Filtering: Recognizing the potential for noise introduction inherent in direct KD, we strategically implement PL filtering to refine the dataset beforehand. This preprocessing step involves utilizing a teacher model to generate predictions for unannotated images. We only include images with predicted scores above a predefined confidence threshold as additional training data. By incorporating these high-confidence predictions, we limit the introduction of noise from incorrect labels.

We implement a dual-loss strategy, combining supervised loss and KD loss. KD loss is computed on all the selected data, while the supervised loss is only computed on data that has annotations. The combined loss, balanced by a constant parameter α, is designed to optimize learning from both annotated data and the nuanced insights provided by the teacher model through soft labels:

$$\mathcal{L} = \begin{cases} \text{for annotated or weakly-annotated data if } f(x;\theta_t) > T, \\ \quad \mathcal{L}_{\text{Sup}}(y, f(\tilde{x};\theta)) + \alpha \cdot \mathcal{L}_{\text{KD}}(f(\tilde{x};\theta_t), f(\tilde{x};\theta)) \\ \text{for unannotated data if } f(x;\theta_t) > T, \\ \quad \alpha \cdot \mathcal{L}_{\text{KD}}(f(\tilde{x};\theta_t), f(\tilde{x};\theta)) \end{cases} \quad (1)$$

where y denotes the annotations; $f(\tilde{x}; \theta)$ represents the predictions for the augmented image $\tilde{x}$ using the student model parameters θ; $f(\tilde{x}; \theta_t)$ and $f(x; \theta_t)$ are the soft labels from the teacher model θ_t on the augmented and non augmented image respectively; and T represents the confidence threshold. We utilize binary cross-entropy loss for the supervised loss ($\mathcal{L}_{\text{Sup}}$) and mean-square error loss for the KD loss ($\mathcal{L}_{\text{KD}}$). This formulation is uniformly applied across both classification and segmentation tasks. Only the KD loss is applied for weakly annotated data in segmentation, as direct annotations are unavailable. This approach ensures that the model learns effectively from both direct annotations and the distilled insights of the teacher model.

3 Experiments

3.1 Dataset

Our proprietary DBT dataset encompasses a total of 13,150 four-view DBT exams from multiple U.S. institutions. The dataset includes 2,487 cancer-positive exams confirmed by biopsy, 3,398 normal exams, and 7,265 benign findings, of which 2,773 are biopsy-confirmed, and 4,492 have a follow-up period of at least one year confirming benign status. The dataset comprises 6,823 exams from recorded with a Hologic device, 2,353 exams from GE and 3,974 exams from a Siemens device. All exams recorded with Hologic devices are annotated, containing both breast level and slice level annotations. The GE and Siemens data are only weakly annotated and contain only breast level annotations.

The dataset was split into training (9,174 exams), validation (1,988 exams), and test (1,988 exams) sets. The allocation was performed, ensuring a balanced representation of cancer, benign, and normal exams across each set.

Our datasets are de-identified in strict adherence to the HIPAA Safe Harbor guidelines. This process involves the removal of all Protected Health Information (PHI) from both the image pixels and the DICOM header. Because data is de-identified Institutional Review Board (IRB) approval is not needed.

3.2 Implementation Details

Our model, developed on Pytorch, was distributed over eight NVIDIA V100 GPUs for training. Following the training protocol, we adopted the stochastic gradient descent (SGD) optimizer. We trained for 60,000 iterations with an initial learning rate of 0.012, weight decay of 0.0001, and momentum of 0.9, employing a cosine annealing strategy for learning rate adjustment [15]. Image inputs were resized to 896 × 640 pixels and processed in batches of 192. Data augmentation included geometric modifications such as horizontal and vertical flipping and rotations, along with photometric transformations like brightness and contrast adjustments, Gaussian noise, sharpening, CLAHE, and solarization. Our architecture is built upon a ResNet-34 backbone, pre-trained on ImageNet, serving as the foundation for both the teacher and student models in our semi-supervised learning setup. The confidence thresholds pseudo-labeling (PL) sampling was selected through grid search and fixed to 0.7 when using weak annotations and 0.1 without. The stride for sampling slices in cancer-negative DBT stacks is 2. For the weighting of the KD loss, we set $\alpha_{\text{seg}} = 25$ for segmentation and $\alpha_{\text{clf}} = 1$ for classification.

3.3 Experimental Setting

We first trained a model using only annotated exams to serve as the teacher model for our semi-supervised experiments with knowledge distillation (KD) and provide a baseline for comparison. For the semi-supervised experiments, we considered several strategies for selecting slices from the weakly annotated datasets. The different experiment settings are as follows:

1. **Baseline**: In this setting, we only use the annotated dataset for training, and KD is not used. This baseline is used as the teacher model for the semi-supervised experiments.
2. **KD**: In this setting, we use KD and also include weakly annotated data in training but ignore weak annotations, therefore considering the added data as unannotated.
3. **KD with weak annotations**: In this setting, we use KD and include weakly annotated data in training. For cancer-positive breasts, we select all DBT slices. Although we know the breast is cancer-positive, we don't know which exact slices contain the lesion, so only KD loss is computed.
4. **SelectiveKD**: This setting is the same as KD, but we also include pseudo-labeling filtering.
5. **SelectiveKD with weak annotations**: This setting is the same as KD with weak annotations. However, we also include pseudo-labeling filtering and use supervised and KD loss on selected slices.

3.4 Results

Models are compared using the Area Under the Receiver Operating Characteristic Curve (AUC). To compute confidence intervals and compare models, we utilized the DeLong test [7]. Results are shown in Table 1.

Efficiency of SelectiveKD. SelectiveKD outperformed the baseline with statistical significance ($p < 0.001$) both with and without using weak annotations, and it also consistently outperformed KD, although the difference was found to be statistically significant (p < 0.001) only when using weak annotations. The best overall performance is achieved with SelectiveKD using weak annotations. We believe there are three main reasons for this; one is that weak annotations also allow the computation of supervised loss for the classification task, adding extra supervision to the model. Second, using weak annotations further mitigates the potential noise introduced by the teacher model by preventing the inclusion of false positive slices from cancer-negative exams, as the inclusion of slices based

Table 1. Main results: the table shows the ROC AUC and the associated 95% confidence intervals for various ablations of our method. We evaluate on Hologic, Siemens, and GE test splits as well as all 3 together. The asterisk (*) indicates models that incorporated weakly annotated data. We highlight that annotations are only available for cases from the Hologic subset and our approach contributes to especially large performance gains in subsets without annotations (Siemens, GE).

Experiment	Hologic		Siemens		GE		All	
	AUC	Conf. Int.	AUC	Conf. Int.	AUC	Conf. Int.	AUC	Conf. Int.
Baseline	0.872	[0.849, 0.895]	0.780	[0.720, 0.841]	0.863	[0.811, 0.916]	0.855	[0.834, 0.876]
KD	0.882	[0.860, 0.904]	0.806	[0.748, 0.865]	0.882	[0.838, 0.926]	0.870	[0.850, 0.889]
SelectiveKD	0.883	[0.860, 0.906]	0.824	[0.769, 0.879]	0.892	[0.846, 0.938]	0.877	[0.858, 0.896]
KD*	0.880	[0.857, 0.902]	0.834	[0.776, 0.891]	0.862	[0.808, 0.916]	0.862	[0.842, 0.883]
SelectiveKD*	0.891	[0.870, 0.912]	0.862	[0.814, 0.909]	0.889	[0.841, 0.937]	0.879	[0.860, 0.897]

on model predictions is conditioned on the breast label being positive. Finally, the optimal confidence threshold when using weak annotations was found to be lower than that without using weak annotations (0.1 and 0.7, respectively). This allows the inclusion of more slices in training and makes sense considering that with weak annotations, the model is applied only on cancer-positive exams, while without them, it is applied on both cancer-positive and negative exams; therefore, it requires a higher confidence threshold to filter out false positive slices from cancer-negative exams.

Generalization. Interestingly, the biggest improvements in performance were obtained on the Siemens and GE datasets, which can be considered as out-of-domain distribution for the baseline since it is trained only on Hologic data. This shows our method's potential to improve generalization capability. The fact that SelectiveKD consistently outperformed KD on Siemens and GE and that the gap between the two methods is larger for Siemens and GE datasets compared to Hologic provides evidence that the use of PL filtering can mitigate the noise introduced by wrong teacher model predictions, especially when incorporating data that is out of distribution for the teacher model.

3.5 Annotation Cost Efficiency

The previous results show that our semi-supervised framework can be used to leverage unannotated data and reach better performance and generalization.

In this part, we focus on analyzing how the performance scales with respect to the proportion of annotated data. To that end, we used our annotated Hologic dataset and ran multiple experiments to gradually increase the proportion of annotated exams included in training from 10 to 100% with 10% increments. For each 10% increment, we create a corresponding dataset by randomly sampling 10% of cancer-positive, benign, and cancer-negative exams, respectively, to ensure the proportion of each subgroup remains constant, which we add to the set of previously sampled annotated exams while the annotations for remaining exams are ignored. In total, 10 different datasets were created that contain the same exams but only the presence of annotation changes.

In this series of experiments, we first trained a baseline model on each dataset to serve as the teacher model for the semi-supervised setting and provide a baseline for comparison. We then applied the SelectiveKD and SelectiveKD with weak annotation settings on each dataset.

As shown in Fig. 2, our semi-supervised framework can offer meaningful performance improvements, especially when only a small portion of the data is annotated. Using only 20% of annotated exams, SelectiveKD is able to match the performance of the baseline using 60% of annotated exams and that of the baseline using 70% of annotated exams when also using weak annotations. As expected, the gap in performance between all 3 settings decreases as the proportion of annotated exams gets closer to 100%. Detailed results, including confidence intervals, can be found in the Supplementary Materials.

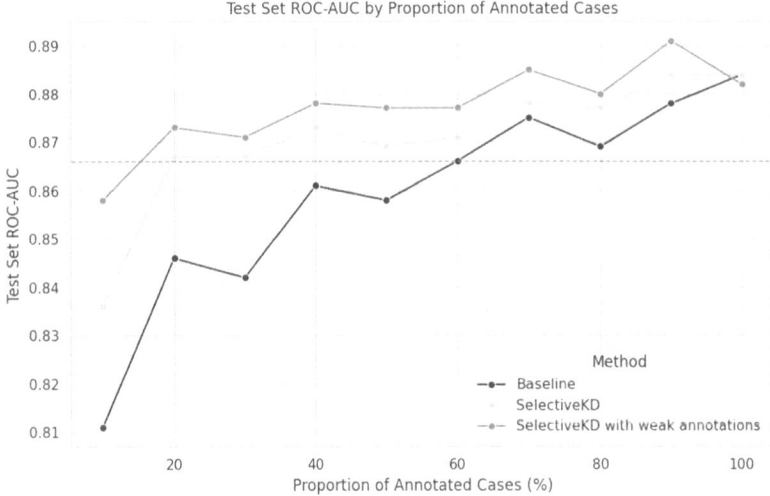

Fig. 2. Test set ROC AUC values versus the proportion of annotated exams: The scaling of model performance with respect to the proportion of annotated data is illustrated for the baseline and SelectiveKD. By using our approach, we show that the model achieves similar level of cancer detection performance while using less than one third of annotations (dotted line).

4 Conclusion

We introduced SelectiveKD, a framework for building a cancer detection CAD for DBT that leverages semi-supervised learning to utilize unannotated slices from DBT stacks. By leveraging knowledge distillation (KD) and selective dataset expansion through pseudo-label (PL) filtering, we showed that we can significantly improve cancer detection performance on a large-scale dataset of over 10,000 DBT exams. Further, we demonstrated that our approach contributes to improving the model's ability to generalize without the need to explicitly collect annotations for each device manufacturer. Finally, our findings suggest that we can build powerful DBT CAD models with fewer annotations by adding more unannotated or weakly annotated cases.

As limitations of this study, the presented framework mainly focused on improving the classification performance of CAD. We believe a deeper investigation of how our approach impacts lesion localization performance as well will add more clinical value to the model. Further, more thorough investigation of CAD model performance across various patient subgroups across multiple institutions is warranted to build a stronger evidence for the efficacy of the presented framework for real world use. We believe the presented framework and the findings from this study contribute toward increasing the practical value of deep learning based CAD for improving breast cancer screening using DBT.

References

1. Andharia, D., Shah, H., Prajapati, A.D., Bhansali, A.D., Shah, A., Desai, D.: Digital breast tomosynthesis(DBT) vs 2D mammography and impact of combined use: a meta-analysis. In: medRxiv (2023)
2. Bahl, M., Gaffney, S., McCarthy, A.M., Lowry, K.P., Dang, P.A., Lehman, C.D.: Breast cancer characteristics associated with 2d digital mammography versus digital breast tomosynthesis for screening-detected and interval cancers. Radiology **287**(1), 49–57 (2018)
3. Berthelot, D., Carlini, N., Goodfellow, I., Papernot, N., Oliver, A., Raffel, C.A.: MixMatch: a holistic approach to semi-supervised learning. Adv. Neural Inf. Process. Syst. **32** (2019)
4. Chen, Y., Mancini, M., Zhu, X., Akata, Z.: Semi-supervised and unsupervised deep visual learning: a survey. IEEE Trans. Pattern Anal. Mach. Intell. **46**, 1327–1347 (2022)
5. Chen, Y., Zhu, X., Li, W., Gong, S.: Semi-supervised learning under class distribution mismatch. In: Proceedings of the AAAI Conference on Artificial Intelligence, vol. 34, pp. 3569–3576 (2020)
6. Chikarmane, S.A., Cochon, L.R., Khorasani, R., Sahu, S., Giess, C.S.: Screening mammography performance metrics of 2D digital mammography versus digital breast tomosynthesis in women with a personal history of breast cancer. Am. J. Roentgenol. **217**, 587–594 (2021)
7. DeLong, E.R., DeLong, D.M., Clarke-Pearson, D.L.: Comparing the areas under two or more correlated receiver operating characteristic curves: a nonparametric approach. Biometrics **44**, 837–845 (1988)
8. Dhamija, E., Gulati, M., Deo, S., Gogia, A., Hari, S.: Digital breast tomosynthesis: an overview. Indian J. Surg. Oncol. **12**(2), 315–329 (2021)
9. Goldberg, J.E., et al.: New horizons: artificial intelligence for digital breast tomosynthesis. Radiographics **43**(1), e220060 (2022)
10. Hinton, G., Vinyals, O., Dean, J.: Distilling the knowledge in a neural network (2015)

11. Huang, L., Chen, Y., He, X.: Spectral-spatial masked transformer with supervised and contrastive learning for hyperspectral image classification. IEEE Trans. Geosci. Remote Sens. **61**, 1–18 (2023)
12. Ko, M.J., et al.: Accuracy of digital breast tomosynthesis for detecting breast cancer in the diagnostic setting: a systematic review and meta-analysis. Korean J. Radiol. **22**(8), 1240 (2021)
13. Lee, D.H., et al.: Pseudo-label: the simple and efficient semi-supervised learning method for deep neural networks. In: Workshop on Challenges in Representation Learning, ICML, Atlanta, vol. 3, p. 896 (2013)
14. Lee, W., Lee, H., Lee, H., Park, E.K., Nam, H., Kooi, T.: Transformer-based deep neural network for breast cancer classification on digital breast tomosynthesis images. Radiol. Artif. Intell. **5**(3), e220159 (2023)
15. Loshchilov, I., Hutter, F.: SGDR: stochastic gradient descent with warm restarts. arXiv preprint arXiv:1608.03983 (2016)
16. Nguyen, T., et al.: Overview of digital breast tomosynthesis: clinical cases, benefits, and disadvantages. Diagn. Interv. Imaging **96**(9), 843–859 (2015)
17. Shahan, C.L.: An overview of digital breast tomosynthesis. W. Va. Med. J. **2016**, 996 (2016)
18. Shen, Y., Shi, L., Zhao, J., Dong, Y., Wang, L.: Fully convolutional spectral-spatial fusion network integrating supervised contrastive learning for hyperspectral image classification. IEEE J. Sel. Topics Appl. Earth Obs. Rem. Sens. **16**, 9077–9088 (2023)
19. Shoshan, Y., et al.: Artificial intelligence for reducing workload in breast cancer screening with digital breast tomosynthesis. Radiology **303**(1), 69–77 (2022)
20. Singh, P., et al.: Shifting to machine supervision: annotation-efficient semi and self-supervised learning for automatic medical image segmentation and classification. arXiv preprint arXiv:2311.10319 (2023)
21. Tamé, I.d.A., Sirotkin, K., Carballeira, P., Escudero-Viñolo, M.: Self-supervised curricular deep learning for chest X-ray image classification. arXiv preprint arXiv:2301.10687 (2023)
22. Tardy, M., Mateus, D.: Trainable summarization to improve breast tomosynthesis classification. In: de Bruijne, M. (ed.) MICCAI 2021. LNCS, vol. 12907, pp. 140–149. Springer, Cham (2021). https://doi.org/10.1007/978 3 030-87234-2_14
23. Xie, Q., Luong, M.T., Hovy, E., Le, Q.V.: Self-training with noisy student improves imageNet classification. In: Proceedings of the IEEE/CVF Conference on Computer Vision and Pattern Recognition, pp. 10687–10698 (2020)
24. Zhang, J., Yang, J., Yu, J., Fan, J.: Semisupervised image classification by mutual learning of multiple self-supervised models. Int. J. Intell. Syst. **37**(5), 3117–3141 (2022)

# Cancer/Early Cancer Detection, Treatment, and Survival Prognosis

AI Age Discrepancy: A Novel Parameter for Frailty Assessment in Kidney Tumor Patients

Rikhil Seshadri[1], Jayant Siva[1], Angelica Bartholomew[1], Clara Goebel[1], Gabriel Wallerstein-King[1], Beatriz López Morato[1], Nicholas Heller[1(✉)], Jason Scovell[1], Rebecca Campbell[1], Andrew Wood[1], Michal Ozery-Flato[2], Vesna Barros[2], Maria Gabrani[3], Michal Rosen-Zvi[2], Resha Tejpaul[4], Vidhyalakshmi Ramesh[4], Nikolaos Papanikolopoulos[4], Subodh Regmi[4], Ryan Ward[1], Robert Abouassaly[1], Steven C. Campbell[1], Erick Remer[1], and Christopher Weight[1]

[1] Cleveland Clinic, Cleveland, USA
rseshadri@gatech.edu, jms665@case.edu,
hellern@ccf.org
[2] IBM Research, Haifa, Israel
[3] IBM Research, Zurich, Switzerland
[4] University of Minnesota, Minneapolis, USA

Abstract. Kidney cancer is a global health concern, and accurate assessment of patient frailty is crucial for optimizing surgical outcomes. This paper introduces AI Age Discrepancy, a novel metric derived from machine learning analysis of preoperative abdominal CT scans, as a potential indicator of frailty and postoperative risk in kidney cancer patients. This retrospective study of 599 patients from the 2023 Kidney Tumor Segmentation (KiTS) challenge dataset found that a higher AI Age Discrepancy is significantly associated with longer hospital stays and lower overall survival rates, independent of established factors. This suggests that AI Age Discrepancy may provide valuable insights into patient frailty and could thus inform clinical decision-making in kidney cancer treatment.

Keywords: Kidney Cancer · Machine Learning · Frailty

1 Introduction

Kidney cancer is a malignancy that originates from the tissues of the kidney or renal pelvis and is primarily discovered through the presence of a kidney tumor on abdominal imaging. While some kidney tumors are benign and do not carry a risk for metastasis, most are malignant [6]. In 2020, kidney cancer made up 2.2% of all cancer diagnoses and 1.8% of all cancer-related fatalities globally. Kidney cancer poses a substantial global health burden with an age-standardized incidence rate of 6.1 per 100,000 in men [15].

R. Seshadri and J. Siva—These authors contributed equally.

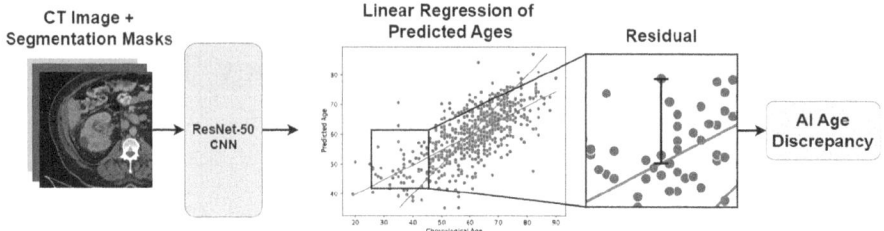

Fig. 1. Proposed AI Age Discrepancy Process. ResNet-50 convolutional neural networks are trained to predict age from preoperative CT images. The linear regression line of predictions is calculated. Residuals from predicated age to regression line are taken and then normalized to obtain the AI Age Discrepancy.

Once a kidney tumor has been identified on imaging, current treatments involve various surgical approaches, each with distinct advantages and challenges. Open nephrectomy, which uses an incision to remove the kidney in its entirety, provides direct access into the retroperitoneum but results in longer recovery times and higher complication risks [12]. In contrast, minimally invasive nephrectomy, including robotic-assisted techniques, uses small incisions and a camera, often resulting in shorter hospital stays, reduced pain, and faster recovery. However, these minimally invasive procedures require specialized skills and may not be as widely available as other interventions [12]. Cryoablation is another minimally invasive procedure used for small, early-stage renal tumors. This technique involves inserting a probe into the tumor to freeze and destroy cancer cells under CT imaging guidance. Cryoablation is best suited to patients who cannot undergo surgery, as it carries a higher risk of local recurrence when compared to surgical options [3]. Nephron-sparing procedures, including partial nephrectomy and cryoablation, are often preferred over radical nephrectomy for smaller renal tumors because they better preserve long-term renal function and lower non-cancer-related mortality rates. Despite being the current gold standard, partial nephrectomy is technically more challenging and associated with a higher rate of minor complications compared to radical nephrectomy [9].

Tumor stage, grade, and surgical approach are well-established factors influencing both length of hospital stay (LOS) and overall survival (OS) after kidney tumor surgery [4] [12]. LOS in this case is defined as the amount of time a patient spends in a hospital or healthcare facility after kidney tumor surgery. OS in this case is defined as the length of time from the date of kidney tumor surgery to the date of death. The AUA guidelines delineate tumor stage, grade, and histology as primary determinants of post-intervention prognosis [4]. Upstaging during surgery (e.g., cT1 to pT3a) indicates a poorer prognosis, highlighting the importance of accurate preoperative staging [16]. Surgical approaches like laparoscopy and robotic surgery also offer faster recovery compared to open nephrectomy [12]. Choosing the appropriate surgical approach depends on tumor size, surgeon expertise, and, the overall health status or 'frailty' of the patient in question. These factors can play a significant role in patients' LOS and OS.

Previous research has found that deep neural networks can be trained to predict a patient's age from CT scans. For instance, Kerber et al. (2023) proposed a deep learning model for age estimation from thoracic and abdominal CT scans, emphasizing the importance of prediction reliability and identifying regions significant for age estimation [11]. Azarfar et al. (2023) also developed a model for age estimation from chest CT scans, suggesting that CT-estimated age may be a useful addition to the calculation of cancer risk [1]. These prior works fall into the emerging category of "biological age" quantification [2] which aims to use biological measurements to quantify age-related changes in a patient's health.

The present study aims to build on these findings by considering AI Age Discrepancy as a predictor of frailty in the context of kidney tumor treatment. This novel parameter may provide a valuable independent signal about patients' overall health status to complement established frailty indices such as the Charlson Comorbidity Index (CCI) [14].

2 Methods

2.1 Dataset

This retrospective study included 599 patients who underwent treatment for suspected renal malignancy at a single institution from 2010 to 2021. The patients were taken from the publicly available 2023 Kidney Tumor Segmentation (KiTS) challenge held in conjunction with the International Conference on Medical Image Computing and Computer-Assisted Interventions (MICCAI) in 2023 [8]. 9 patients under the age of 18 were excluded since the mechanisms for age-related frailty are not likely to impact adolescents. All patients underwent preoperative contrast-enhanced abdominal CT imaging in either the arterial or venous phase. The original compilation and public release of the KiTS23 cohort received approval from the University of Minnesota Institutional Review Board. This subsequent analysis of de-identified data was deemed exempt from IRB review.

2.2 Age Prediction Model

A ResNet-50 architecture with pre-trained weights from the ImageNet database was fine-tuned to predict age as a continuous variable. 2-dimensional views were extracted from all three anatomical planes, and three channels were constructed containing (1) the CT image Hounsfield Unit attenuation values, (2) a mask representing the tumor segmentation, and (3) a mask representing the affected kidney. Scans in the training set were randomly sampled for training, and 2D views were sampled from selected scans with weighted probabilities corresponding to the fraction of tumor voxels in each view. Final age predictions were obtained by computing a weighted average of the predictions made on all views, with weight again determined by fraction of tumor voxels.

5-fold cross-validation was repeated three times to obtain average test-set age predictions for the full dataset. A least squares regression fit between the predicted age and the chronological age of the patient is shown in Fig. 2. Some regression to the mean is observed, and must be adjusted for in downstream analyses. Therefore, the residuals between the predicted age and the regression line (as opposed to y=x) were calculated, and the residuals were then normalized with a mean of 0 and a standard deviation of 1. We refer to these normalized residuals as the patients' AI Age Discrepancy. We hypothesize that the AI Age Discrepancy captures a measure of a patient's biological health, potentially reflecting underlying health status that may not be fully revealed by chronological age. This makes intuitive sense because a model trained to predict age will learn to identify age-related changes, but these changes could occur more rapidly in some patients than in others. Hence, a positive AI Age Discrepancy reflects that a patient is predicted to be older than expected, which could reflect poorer biological health than is typical for their age.

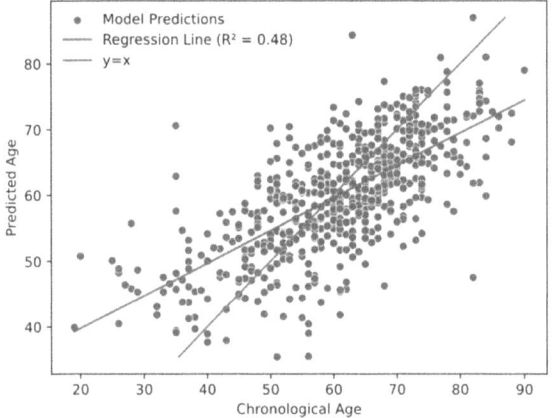

Fig. 2. Scatter plot of patient predicted ages (blue), linear regression line of predicted ages (red), and line of perfect predictions (grey). AI Age Discrepancy is the normalized residual from the predicted age to the linear regression line. (Color figure online)

2.3 Statistical Tests

Multivariate Cox proportional-hazards regression was used to evaluate the association between AI Age Discrepancy and our two endpoints, LOS and OS, and the models included established covariates for each outcome. For LOS, the covariates include surgical approach (laparoscopic/robotic vs open nephrectomy), non-nephron sparing vs nephron-sparing surgery, CCI points, tumor size, and chronological age. For OS, additional covariates included tumor stage (greater than or equal to 3 or less than 3), lymph node involvement, metastasis, and ordinal ISUP grade. Stage and grade were not included in hospital stay predictions as they are not known before surgical intervention. The Lifelines Python library was used for statistical analysis.

3 Results

We analyzed data from 590 patients who underwent surgical intervention for kidney cancer between 2010 and 2021 using Cox proportional hazards regression to identify factors associated with LOS and OS. Figure 3 summarizes the Cox regression analysis of AI Age Discrepancy as a predictor of LOS and OS in a multivariate model.

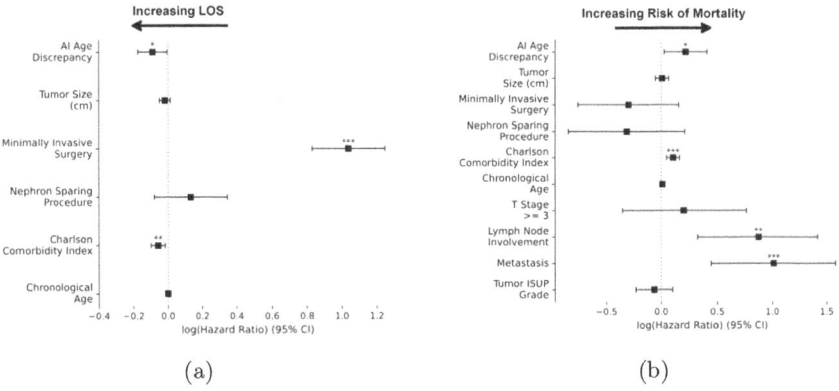

Fig. 3. Forest plot analysis of (a) LOS and (b) OS. This summarizes the results of the Cox proportional hazards regression to identify factors associated with LOS and OS following kidney cancer surgery. The log(HR) and 95% confidence intervals are presented for each variable. (a) Lower log(HR) values indicate factors associated with a longer LOS. (b) Higher log(HR) values indicate factors associated with an increasing risk of mortality.

3.1 Length of Hospital Stay

Our analysis revealed a statistically significant association between AI Age Discrepancy and LOS. Table 1 summarizes the results of the Cox regression analysis of variables associated with LOS. The hazard ratio (HR) for AI Age Discrepancy was 0.914 (95% CI: 0.840 to 0.994, p=0.036). Minimally invasive surgery was also associated with a significantly shorter LOS when compared to other surgical approaches with a HR of 2.825 (95% CI: 2.293 to 3.480, p<0.001). As expected, higher Charlson Comorbidity Index (CCI) scores, which reflect a greater burden of comorbidities, were associated with a lower likelihood of early discharge (HR=0.944, 95% CI: 0.905 to 0.983, p=0.006). The presence of a statistically significant association between LOS and our proposed AI Age Discrepancy independent of these known predictors demonstrates the potential of our approach to provide complementary information.

Table 1. Cox regression analysis of LOS. This summarizes the results of the Cox proportional hazards regression to identify variables associated with LOS following kidney cancer surgery.

Variable	Hazard Ratio	95% CI	p-value
AI Age Discrepancy	0.914	(0.840–0.994)	**0.036**
Tumor Size	0.981	(0.950–1.013)	0.240
Minimally Invasive Surgery	2.825	(2.293–3.480)	**<0.001**
Nephron Sparing Procedure	1.141	(0.922–1.412)	0.225
Charlson Comorbidity Index	0.944	(0.905–0.983)	**0.006**
Chronological Age	1.000	(0.992–1.008)	0.994

3.2 Overall Survival

Our analysis also revealed a statistically significant association between AI Age Discrepancy and OS. Table 2 summarizes the results of the Cox regression analysis of variables associated with OS. The HR for AI Age Discrepancy was 1.242 (95% CI: 1.025 to 1.504, p=0.027). An HR greater than 1 indicates that a higher AI Age Discrepancy is associated with a lower likelihood of longer OS. Lymph node involvement also emerged as a significant predictor of OS (HR=2.388, 95% CI: 1.384 to 4.120, p=0.002). Metastasis, as expected, was associated with worse prognosis (HR=2.753, 95% CI: 1.562 to 4.848, p<0.001). Like with LOS, increased CCI points were associated with a worse prognosis (HR=1.112, 95% CI: 1.049 to 1.180, p<0.001).

Table 2. Cox regression analysis of OS. This summarizes the results of the Cox proportional hazards regression to identify variables associated with OS following kidney cancer surgery.

Variable	Hazard Ratio	95% CI	p-value
AI Age Discrepancy	1.242	(1.025–1.504)	**0.027**
Tumor Size	1.007	(0.947–1.071)	0.814
Minimally Invasive Surgery	0.741	(0.469–1.170)	0.198
Nephron Sparing Procedure	0.729	(0.430–1.234)	0.239
Charlson Comorbidity Index	1.112	(1.049–1.180)	**<0.001**
Chronological Age	1.011	(0.991–1.031)	0.275
T stage ≥ 3	1.223	(0.701–2.134)	0.479
Lymph Node Involvement	2.388	(1.384–4.120)	**0.002**
Metastasis	2.753	(1.562–4.848)	**<0.001**
Tumor ISUP Grade	0.938	(0.793–1.109)	0.454

4 Discussion

This study introduces AI Age Discrepancy, a novel parameter for quantifying frailty in patients undergoing surgery for kidney cancer. This approach is based on the observation that deep neural networks can learn to predict patient age with reasonable accuracy from only an abdominal pre-operative CT scan [11]. We hypothesized that patients whose AI Age Discrepancy is higher would be at a greater risk for poor postoperative outcomes following abdominal surgery like partial or radical nephrectomy.

Our analysis reveals a statistically significant association between AI Age Discrepancy and LOS. Patients with a higher AI Age Discrepancy had a longer LOS (HR < 1). While this study is observational and causal relationships cannot be established, the findings suggest that AI Age Discrepancy could be a valuable variable to consider when making decisions regarding the surgical management of renal masses. Future prospective studies are needed to validate the clinical utility of AI Age Discrepancy for predicting postoperative outcomes in comparison to current metrics, such as the Charlson Comorbidity Index (CCI).

One possible explanation for the accuracy of our machine learning model in estimating patient age is the consideration of sarcopenia, a condition characterized by the loss of skeletal muscle mass and function. Sarcopenia is associated with aging and various chronic conditions, impacting overall health and physical performance [13] [7]. Multiple metabolic processes and physical abilities are impacted by the loss of skeletal muscle mass, serving as a potential indicator of biological aging [17]. CT imaging, particularly the assessment of muscle areas such as the erector spinae at the T12 vertebra, provides a reliable measure of muscle mass and can reflect sarcopenia severity [5].

In addition to sarcopenia, other anatomical and physiological features captured on CT imaging may contribute to the model's accuracy. For instance, the vertebral column can provide insights into degenerative changes and bone density, both of which correlate with aging as a marker of osteopenia or osteoporosis. The development of osteophytes on the vertebral column is a common age-related change and can be accurately assessed using deep learning algorithms, improving age estimation significantly [10]. The volume of parenchymal tissue, which includes the functional parts of organs such as the kidneys, can also indicate age-related changes. Furthermore, visceral fat, which accumulates around internal organs, is known to increase with age and is associated with various metabolic diseases [5].

We postulate that by integrating these diverse data points from CT scans, our machine-learning model can capture a comprehensive picture of a patient's biological age and health status. This multifaceted approach potentially enhances the model's predictive power for both age and associated comorbidities, ultimately improving the accuracy of post-surgical outcome predictions. The survival analysis identified established factors like lymph node involvement and metastasis as significant predictors of post-surgical survival. This aligns with existing knowledge about the association between these factors and a poorer prognosis in kidney cancer patients [4]. Importantly, our proposed AI Age Discrepancy was a statistically significant predictor independent of these factors.

This study demonstrates the potential of deep learning and CT-derived information for clinical decision-making in kidney cancer surgery. We anticipate that the integration of AI Age Discrepancy as a predictor of frailty into clinical decision-making processes could play a significant role in improving patient outcomes, including shared-decision making regarding surgical approach as well as anticipating higher risks for complications and, thus, closer preventative monitoring of patients with higher AI Age Discrepancies. This variable might be useful for clinicians when deciding which patients to operate on and how aggressive a surgery they might tolerate. Existing frailty assessments, such as the Clinical Frailty Scale (CFS) and Modified Frailty Index (mFI) are limited in their clinical value in the surgical sphere, as they are subject to variability due to subjectivity in interpretations of the variables that comprise each respective assessment. AI Age Discrepancy as a parameter appears to be less susceptible to variability and subjectivity in nature. Future studies on how to successfully integrate this metric might reflect a significant clinical benefit to clinicians and patients alike in shared decision-making processes.

Future research is needed to externally validate our retrospective findings, and the clinical utility of AI Age Discrepancy could be evaluated using prospective studies in which predicted age information is considered during treatment planning. Additionally, investigating the biological mechanisms underlying the association between AI Age Discrepancy and outcomes could carry significant implications once further elucidated. Finally, given the broad utility of assessing patient frailty, similar approaches may also prove useful for other surgical procedures.

5 Conclusion

This study presents a novel parameter, AI Age Discrepancy, as a method for quantifying the frailty and postoperative risk associated with kidney tumor surgery. Our findings reveal that a higher AI Age Discrepancy is associated with both longer hospital stays and shorter overall survival. The deep network's accuracy in estimating age likely results from its ability to capture age-related features on CT scans, such as muscle mass, bone density, and organ volume. By integrating this information, the model quantifies a more comprehensive picture of a patient's biological health, which may be useful in objectively stratifying the risk associated with invasive procedures such as partial and radical nephrectomy. This study highlights the potential for deep learning and CT-derived information to improve surgical decision-making in kidney cancer. Future research is needed to validate these findings on an external cohort, and prospective studies could be conducted to more thoroughly probe the clinical utility of considering this variable in decisions for kidney tumor care.

References

1. Azarfar, G., Ko, S.-B.S.-B., Adams, S.J., Babyn, P.S.: Deep learning-based age estimation from chest CT scans. Int. J. Comput. Assist. Radiol. Surg. **19**, 119–127 (2023)
2. Babyn, P.S., Adams, S.J.: Ai analysis of chest radiographs as a biomarker of biological age. The Lancet Healthy Longevity **4**(9), e446–e447 (2023)
3. Bisbee, C.A., Zhang, J., Owens, J., Hussain, S.: Cryoablation for the treatment of kidney cancer: comparison with other treatment modalities and review of current treatment. Cureus (2022)
4. Campbell, S.C., Uzzo, R.G., Karam, J.A., Chang, S.S., Clark, P.E., Souter, L.: Renal mass and localized renal cancer: Evaluation, management, and follow-up: aua guideline: part ii, 8 (2021)
5. Cao, J., et al.: Correlation between bioelectrical impedance analysis and chest CT-measured erector spinae muscle area: a cross-sectional study. Front. Endocrinol. **13**, 7 (2022)
6. Chawla, S.N., Crispen, P.L., Hanlon, A.L., Greenberg, R.E., Chen, D.Y.T., Uzzo, R.G.: Meta-analysis and review of the world literature: the natural history of observed enhancing renal masses. J. Urol. **175**, 425–431 (2006)
7. Dutta, C., Hadley, E.C., Lexell, J.: Sarcopenia and physical performance in old age: overview (1997)
8. Nicholas Heller, et al.: The KiTS21 challenge: automatic segmentation of kidneys, renal tumors, and renal cysts in corticomedullary-phase CT (2023)
9. Kalogirou, C., et al.: Long-term outcome of nephron-sparing surgery compared to radical nephrectomy for renal cell carcinoma $>=$4 cm - a matched-pair single institution analysis. Urologia Int. **98**, 138–147 (2017)
10. Kawashita, I., et al.: Development of a deep-learning algorithm for age estimation on CT images of the vertebral column. Leg. Med. **69**, 7 (2024)
11. Kerber, B., Hepp, T., Küstner, T., Gatidis, S.: Deep learning-based age estimation from clinical computed tomography image data of the thorax and abdomen in the adult population. PLOS ONE **18**, e0292993 (2023)
12. Kunath, F., et al.: Partial nephrectomy versus radical nephrectomy for clinical localised renal masses (2017)
13. Sabatino, A., Cuppari, L., Stenvinkel, P., Lindholm, B., Avesani, C.M.: Sarcopenia in chronic kidney disease: what have we learned so far? (2021)
14. Sundararajan, V., Henderson, T., Perry, C., Muggivan, A., Quan, H., Ghali, W.A.: New ICD-10 version of the Charlson comorbidity index predicted in-hospital mortality. J. Clin. Epidemiol. **57**, 1288–1294 (2004)
15. Sung, H., Ferlay, J., Siegel, R.L., Laversanne, M., Soerjomataram, I., Jemal, A., Bray, F.: Global Cancer Statistics 2020: GLOBOCAN Estimates of Incidence and Mortality Worldwide for 36 Cancers in 185 Countries. CA Cancer J. Clin. **71**(3), 209–249 (2021). https://doi.org/10.3322/caac.21660
16. Veccia, A.: Upstaging to pT3a disease in patients undergoing robotic partial nephrectomy for cT1 kidney cancer: outcomes and predictors from a multi-institutional dataset. Urologic Oncol.: Seminars Original Invest. **38**(4), 286–292 (2020). https://doi.org/10.1016/j.urolonc.2019.12.024
17. Wilkinson, D.J., Piasecki, M., Atherton, P.J.: The age-related loss of skeletal muscle mass and function: measurement and physiology of muscle fibre atrophy and muscle fibre loss in humans (2018)

Deep Neural Networks for Predicting Recurrence and Survival in Patients with Esophageal Cancer After Surgery

Yuhan Zheng[1](✉)⃝, Jessie A. Elliott[2], John V. Reynolds[2], Sheraz R. Markar[3], Bartłomiej W. Papież[1](✉)⃝, and ENSURE study group[4]

[1] Big Data Institute, University of Oxford, Oxford, UK
yuhan.zheng@univ.ox.ac.uk, bartlomiej.papiez@bdi.ox.ac.uk
[2] Trinity St. James's Cancer Institute, Trinity College Dublin and St. James's Hospital, Dublin, Ireland
[3] Nuffield Department of Surgery, University of Oxford, Oxford, UK
[4] Young Investigator Division, European Society for Diseases of the Esophagus, Dublin, Ireland

Abstract. Esophageal cancer is a major cause of cancer-related mortality internationally, with high recurrence rates and poor survival even among patients treated with curative-intent surgery. Investigating relevant prognostic factors and predicting prognosis can enhance post-operative clinical decision-making and potentially improve patients' outcomes. In this work, we assessed prognostic factor identification and discriminative performances of three models for Disease-Free Survival (DFS) and Overall Survival (OS) using a large multicenter international dataset from ENSURE study. We first employed Cox Proportional Hazards (CoxPH) model to assess the impact of each feature on outcomes. Subsequently, we utilised CoxPH and two deep neural network (DNN)-based models, DeepSurv and DeepHit, to predict DFS and OS. The significant prognostic factors identified by our models were consistent with clinical literature, with post-operative pathologic features showing higher significance than clinical stage features. DeepSurv and DeepHit demonstrated comparable discriminative accuracy to CoxPH, with DeepSurv slightly outperforming in both DFS and OS prediction tasks, achieving C-index of 0.735 and 0.74, respectively. While these results suggested the potential of DNNs as prognostic tools for improving predictive accuracy and providing personalised guidance with respect to risk stratification, CoxPH still remains an adequately good prediction model, with the data used in this study.

Keywords: Esophageal Cancer · Survival · Recurrence · Deep Neural Networks · Early Intervention · Patient Stratification

---

Supplementary Information The online version contains supplementary material available at https://doi.org/10.1007/978-3-031-73376-5_17.

1 Introduction

Esophageal cancer is a major cause of cancer-related mortality internationally. The average 5-year Overall Survival (OS) rate is less than 25% [1], ranging from 10% to 55% depending on the stage of which the disease is detected [2]. While surgical resection, known as esophagectomy, remains the primary treatment for esophageal cancer, the prognosis of post-operative patients remains poor. Despite advancements in cancer management strategy, more than 50% of the patients experience a recurrence within 1–3 years following curative-intent surgery [3], with a median survival time of 24 months [4]. Therefore, identifying prognostic factors associated with a higher risk of recurrence, as well as predicting and stratifying patients based on their recurrence and survival probabilities, are crucial to the delivery of personalised medicine approaches that could potentially improve oncologic outcomes. Current risk stratification methods for patients with esophageal cancer predominantly rely on pathological data, primarily tumor staging [5]. This does not fully leverage all available clinical and patient-level data efficiently, and does not account for individual variations.

To address these issues, some studies have developed models for prognosis prediction. For example, logistic regression models have been employed to predict absolute risks for patients with esophageal cancer [6,7]. However, these models predict a single-point outcome event without incorporating time-to-event analysis and are limited to one histologic type only. The Cox Proportional Hazards model (CoxPH) [8] is a widely used regression model that allows the study of the relationships between time-to-event outcomes and a set of covariates. Many studies have employed CoxPH to identify prognostic factors for different outcomes [9–11]. However, CoxPH model assumes linear relationships between covariates and that the relative hazard remains constant over time. This hinders its ability to capture higher level interactions between variables and outcomes.

Recent developments in AI have led to increased applications of machine learning (ML) models in oncology to address more complex problems. For example, Zhang et al. [13] explored multiple ML methods for survival prediction in squamous cell carcinoma, and demonstrated that while CoxPH model remains sufficiently good for interpretive studies, ML approaches have the potential to enhance predictive accuracy. Gong et al. [14] explored artificial neural networks (ANNs) in survival prediction, though these did not outperform other traditional ML models such as XGBoost [15]. However, most of these aforementioned studies relied on data collected from a single center, raising questions about their generalisability and robustness when applied to larger multicenter cohorts. Most studies focus on only one type of outcome, and the prediction values on other outcomes remain unknown. Moreover, these studies often utilise a limited number of features. There is a significant clinical interest in incorporating a more comprehensive set of features that take account into, for example, improvements in treatment technologies or surveillance strategies. Gujjuri et al. [12] implemented CoxPH and Random Forest using ENSURE dataset. However, the results showed that Random Forest did not surpass CoxPH in both discrimination and calibration.

In this work, we developed models to predict Disease-Free Survival (DFS) and OS for patients with esophageal cancer following curative-intent surgery. The work is divided into two main components. The first component is prognostic factor identification task, which aims to identify significant prognostic factors that influence outcomes based on their hazard ratios and significance values, thereby providing clinical guidance. The second component is a prediction task, which aims to develop robust models for prognosis prediction on multicenter heterogeneous dataset. This helps stratify patients based on their risks, which could potentially facilitate personalisation of postoperative treatment and surveillance strategies.

Our contributions are threefold. Firstly, we developed models using a large heterogeneous multicenter cohort [16]. Secondly, we incorporated a comprehensive set of easily accessible and readily identifiable features into the models, including several general center-specific features, to explore more broadly prognostic factors. Finally we carried out extensive experiments with deep neural network (DNN)-based models, and compared their predictive performance with CoxPH model.

The remainder of this paper is organised as follows. Section 2 introduces the details of the dataset, preprocessing steps and provides an overview of the final dataset used for this work. The three models employed and the experimental setup which includes training and implementation details, are described in Sect. 3.1 and Sect. 3.2, respectively. Section 4.1 and Sect. 4.2 present results for the prognostic factor identification task and prediction task. Finally, the discussion and conclusion can be found in Sect. 5.

2 Dataset and Preprocessing

2.1 Dataset

This work is based on data collected from the European iNvestigation of SUrveillance after Resection for Esophageal cancer (ENSURE) study [16], a retrospective non-interventional study taken across 20 European centers. Patients with esophageal or junction cancer undergoing curative intent treatment from June 2009 to June 2015 were all considered for inclusion. In total, there are 4972 patients and over 170 variables. All patients were staged according to the 8th edition of the American Joint Committee on Cancer (AJCC) staging [17].

The use of the dataset and this study has been approved by he Joint Research Ethics Committee of Tallaght University Hospital and St. James's Hospital, Dublin, Ireland (SJH-TUH JREC Ref 2943 Amendment 1).

2.2 Outcome Variable Definition

In this work, DFS is defined as the time from treatment (i.e., surgery) to recurrence or death from any cause [18]. Patients who are lost to follow-up or remain alive without recurrence at the end of the study are recorded as censored events.

OS is defined as the period from diagnosis to death from any cause [19]. Patients that are lost to follow-up or still alive at the end of the study are recorded as censored event.

2.3 Patient Inclusion and Variable Selection Criteria

Patient Inclusion. In this work, we removed patients with missing DFS and/or OS outcome, as well as patients with rare histologic type (i.e., non-adenocarcinoma and non-squamous cell carcinoma). We excluded further patients with postoperative death for DFS prediction by definition.

Variable Selection. Variables used in our models were selected by experienced clinicians, based on the literature review and their clinical importance. Variables exhibiting clinically known high correlations with other variables, lacking well-established relationships with outcomes, or variables that were often poorly documented by centers, were excluded from the study. Additionally, while there is no single acceptable threshold for missing rate, the approach to dealing with missingness requires careful consideration. Blindly applying imputation strategies to variables with high missing rate could also impose biases [20]. Therefore, after further assessment by clinicians, a set of variables was additionally removed based on both their rate of missingness and their clinical relevance.

In this work, we did not apply any ML or statistic-based variable selection strategies. Evidence [30] suggests that feature selection prior to model application does not significantly improve model performance, especially that we either adopted regularisers in the model (more details in Sect. 3.2) or the ML models themselves have internal feature selection capabilities to handle high-dimensional data in this study. As a result of this variable selection processes, 37 variables were selected with a missing rate of less than 30%.

2.4 Missingness and Imputation

In this study, the missingness mechanism was assumed to be Missing At Random (MAR) [21], as whether the data is missing or not depends exclusively on their availability at center during data collection process [22,23]. This assumption allows us to apply imputation strategies to handle missingness. A flow chart illustrating the overall process, which is going to described below, can be found in Fig. 1 in Sect. A.1.

Different imputation strategies were applied to the prognostic factor identification task and prediction task that were mentioned in Sect. 1. Multiple Imputation by Chained Equations (MICE) [24] was used for prognostic factor identification task, with 10 iterations per imputation set. Multiple imputation (MI), which takes the uncertainty of imputation into account and fills different multiple plausible values, is important to reduce bias and chance of false-positive and false-negative conclusions [25]. The multiple imputed datasets were passed into models, optimised and analysed separately, and final results were combined using Rubin's rule [26]. For prediction task, where the impact of imputation uncertainty is generally less critical, we used single-point multivariate imputation by chained equations, which is typically sufficient for predictive modeling purposes.

When performing imputation, outcome variables, including the binary event indicator and the time-to-event variable, were also included in the prediction matrix to prevent bias [27]. The time-to-event variable was transformed to its cumulative hazard function with the non-parametric Nelson-Aalen estimator [28] as suggested in [29]. The imputation was conducted within the cross-validation (CV) loop during training [30] to prevent any information leakage from the validation set into the training process.

Prior to imputation, the nominal categorical variables were dummy coded. It is important to note that during the imputation process, continuous values were generated for all dummy-coded binary variables, and these values were not rounded to the nearest integer, as recommended based on the findings in [31]. Additionally, after imputation, continuous numerical variables were scaled by zero-score standardisation to bring all variables to approximately similar dynamic ranges to improve numerical stability during training.

Table 1. Summary of the dataset used for model development. DFS task (n=3921), OS task (n=4077).

Outcome	No. of Variables	No. of Patients	No. of Observed Events	Min.(months)	Max.(months)	Median (months)	Mean (months)
DFS	34	3921	2308	0	173	29.7	36.1
OS	34	4077	2173	0.2	176.7	37.47	41.92

Furthermore, after standardisation, three variables that had Pearson correlation coefficients higher than 70% were removed. While there is no definitive threshold for exclusion, we set this threshold based on the interpretations provided in [32] and common practices in the field. As a result, 34 variables were ultimately selected for model development.

2.5 Data Overview

Table 1 summarises the statistics for the dataset used in DFS and OS tasks, respectively.

3 Methods and Experiments

3.1 Models

In this work, three models were employed to predict DFS and OS: a regression model CoxPH [8] and two neural network-based models named DeepSurv [33] and DeepHit [34]. CoxPH is a semi-parametric regression model that takes the form $h_0(t)exp(\sum_i x_i \cdot \beta_i)$, where $h_0(t)$ is baseline hazard function, x_i is covariate and β_i is coefficient. The model assumes that the effect of a factor is constant over time and there is a linear relationship between predictors and log-hazards. DeepSurv is a DNN-based extension of CoxPH model. It models the hazard function as $h_0(t)exp(f_\theta(x))$, where $f_\theta(x)$ is a neural network that takes covariates as input and outputs a scalar. This allows DeepSurv to capture high-level interactions among features. DeepHit, on the other hand, employs an end-to-end DNN that learns the distribution of survival times directly, without making any assumptions about the underlying stochastic process.

In this work, CoxPH model was employed for the prognostic factor identification task. For the prediction task, all three models were used, with CoxPH serving as a baseline for comparison with neural network-based methods. These models were chosen to leverage their respective strengths in handling different aspects of survival analysis, from traditional regression assumptions to capturing complex interactions and learning distributions directly from data.

3.2 Experimental Setup

Dataset Splitting Strategy. The dataset was split into two parts: 80% for training and 20% as held-out testing dataset. For the prognostic factor identification task, the training set was further split into 85% for training and 15% for validation. Stratified bootstrapping was performed on the validation set to select the best set of hyperparameters. For the prediction task, a stratified 5-fold CV was performed on the 80% training set for hyperparameter selection. The imputation and standardisation were performed within the CV loop to avoid information leakage, as mentioned in Sect. 2.4. A graphical illustration of the splitting strategy can be found in Fig. 2 in Appendix Sect. A.2.

Hyperparameter Tuning. Hyperparameter selection was conducted in a grid-search manner. A detailed list of the optimal set of hyperparameters for each model and task can be found in Table 4 in Appendix Sect. A.3. Elastic net regularisation (i.e., L1 (Lasso) and L2 (Ridge) regularisation penalties) was applied to CoxPH. CoxPH with Elastic net [36] was generally found to outperform standard CoxPH during training.

Performance Evaluation. Three metrics were used to evaluate the discrminative performances of the models: concordance index (C-index), Integrated Brier Score (IBS), and time-dependent AUC (tAUC, also known as dynamic AUC).

Implementation. All the models and analyses were implemented using Python 3.10.5. Survival models were implemented with lifelines 0.28.0 and pycox 0.2.3. The CoxPH was trained on a CPU with a memory of 15.2GB. DeepSurv and DeepHit were trained on NVIDIA GPUs with 40GB of RAM.

Table 2. Multivariate CoxPH analysis results for DFS and OS. Relative hazard ratio was calculated for nominal categorical variables with one category as reference (indicated as 'ref' in the table). P < 0.05 was considered as significant. Only significant variables are listed here. NA: neoadjuvant; CRT: chemoradiation therapy. Definitions and staging criteria of the features can be found in [17].

Variable	DFS		OS	
	HR (95% CI)	P-value	HR (95% CI)	P-value
Sex				
Female	ref			
Male	1.200 (1.199-1.200)	**0.007**	1.134 (1.130-1.138)	0.050
Age (Years)	1.017 (1.001-1.033)	0.553	1.115 (1.114-1.115)	**<0.001**
Clinical N stage				
cN0	ref			
cN2	1.155 (1.147-1.163)	0.081	1.239 (1.238-1.240)	**0.010**
Tumor Site				
Junctional	ref			
Lower	1.207 (1.206-1.207)	**0.006**	1.135 (1.132-1.138)	**0.042**
Middle	1.309 (1.306-1.311)	**0.019**	1.266 (1.262-1.269)	**0.027**
Proximal margin positive	1.994 (1.994-1.994)	**<0.001**	1.292 (1.267-1.318)	0.121
Radial margin positive	1.549 (1.549-1.549)	**<0.001**	1.426 (1.426-1.426)	**<0.001**
Pathologic T stage				
T0	ref			
T3	1.583 (1.583-1.583)	**<0.001**	1.490 (1.490-1.490)	**0.001**
T4	1.672 (1.671-1.672)	**0.002**	1.752 (1.751-1.752)	**0.002**
Pathologic N stage				
N0	ref			
N1	1.484 (1.484-1.484)	**<0.001**	1.245 (1.243-1.246)	**0.013**
N2	1.664 (1.664-1.664)	**<0.001**	1.528 (1.528-1.528)	**<0.001**
N3	3.087 (3.087-3.087)	**<0.001**	2.991 (2.991-2.991)	**<0.001**
Pathologic M stage				
M0	ref			
M1	1.707 (1.707-1.707)	**<0.001**	1.919 (1.919-1.919)	**<0.001**
Differentiation				
Gx, cannot be assessed	ref			
Poorly differentiated	1.379 (1.378-1.379)	**0.004**	1.447 (1.446-1.447)	**0.003**
Lymphatic invasion	1.055 (1.000-1.113)	0.573	1.316 (1.316-1.316)	**<0.001**
Venous invasion	1.292 (1.291-1.292)	**<0.001**	1.086 (1.060-1.112)	0.306
Perineural invasion	1.161 (1.158-1.164)	**0.038**	1.193 (1.193, 1.194)	**0.006**
Number of nodes analyzed	0.894 (0.894, 0.894)	**0.002**	0.883 (0.884, 0.884)	**<0.001**
Treatment protocol				
Surgery only	ref			
NA CRT then surgery	1.326 (1.326-1.326)	**0.003**	1.198 (1.195-1.200)	**0.025**
Clavien-Dindo Grade	1.060 (1.059–1.060)	**<0.001**	1.169 (1.169–1.169)	**<0.001**
Length of stay (Days)	1.077 (1.076–1.077)	**0.010**	1.052 (1.050–1.053)	**0.044**
Cancer cases per year	0.919 (0.919–0.919)	**0.008**	0.925 (0.924–0.926)	**0.024**

4 Results

4.1 Prognostic Factor Identification Task

Table 2 summarises the multivariate analysis results of CoxPH with the significant variables (P-value < 0.05) being listed only, along with their hazard ratio (HR), and 95% confidence interval (CI).

4.2 Prediction Task

Table 3 summarises the discriminative prediction performance of three models for DFS and OS respectively. Comparing all three metrics reveals that DeepSurv demonstrates comparable performances to CoxPH, while DeepHit demonstrates slightly inferior performance in terms of IBS. Figure 3 in Appendix Sect. A.4 provides examples of predicted OS curves obtained from the three models for the same random set of five patients. Notably, while CoxPH and DeepSurv exhibit similar shapes and distributions, DeepHit shows a completely different profile, with minimal variation among the five prediction curves. Despite DeepHit generally ordering patients consistently in terms of survival probabilities compared to the other two models, this profile suggests poorer calibration performance.

Table 3. Summary of model performances. C-index: concordance index; IBS: Integrated Brier Score; tAUC: time-dependent AUC.

	C-index (95% CI) ↑	IBS (95% CI) ↓	tAUC (95% CI) ↑
DFS			
CoxPH	0.733 (0.710, 0.755)	**0.174 (0.160, 0.187)**	0.720 (0.682, 0.799)
DeepSurv	**0.735 (0.714, 0.758)**	0.176 (0.163, 0.193)	**0.749 (0.727, 0.801)**
DeepHit	0.729 (0.707, 0.752)	0.249 (0.243, 0.263)	0.729 (0.693, 0.797)
OS			
CoxPH	0.734 (0.710, 0.758)	**0.164 (0.153, 0.181)**	**0.783 (0.738, 0.818)**
DeepSurv	**0.740 (0.716, 0.764)**	0.169 (0.152, 0.192)	0.781 (0.734, 0.827)
DeepHit	0.739 (0.716, 0.762)	0.214 (0.201, 0.233)	0.776 (0.707, 0.827)

5 Discussion and Conclusion

In this work, we analysed a heterogeneous multicenter dataset to investigate the contribution of covariates to and predictive performance of three models on DFS and OS in patients with esophageal cancer. The significant prognostic factors identified aligned well with clinical literature and experiences. For example, pathologic tumor staging features appear to be strong prognostic factors, and are generally more significant than clinical staging [35]. The more advanced the pathologic stage of the tumor is, the higher the hazard ratio. In terms of prediction, DeepSurv consistently outperformed CoxPH in both DFS and OS tasks, with C-index of 0.735 and 0.740, respectively, when C-index serving as the primary metric. Overall, the two DNN-based models demonstrated comparable discriminative performance to CoxPH; though DeepHit was found to exhibit poorer calibration performance compared to the other two models. The use of a multicenter international dataset, which includes patients with either adenocarcinoma or squamous cell carcinoma, suggested broader applicability of these findings across diverse cohort in various clinical settings. In general, despite their ability to model more complex interactions, DNN-based models did not greatly outperform the CoxPH. The CoxPH, which is interpretable and computationally efficient, still remains a sufficiently good prediction model with tabular data.

While all three models demonstrated good discriminative performance, it is inferred that these results likely represent the upper bound achievable with tabular data. It is worth noting that some significant features, for example, Clinical N stage, are derived from radiologic assessment scans (Computed Tomography (CT), Positron Emission Tomography (PET)) [17]. Therefore, incorporating imaging-derived features such as radiomics could provide more detailed information and potentially enhance model performance [42]. It should be noted that, among the three models, DeepHit posed particular challenges during training, showing large fluctuations in performance and high sensitivity to hyperparameters. This difficulty can be attributed to its end-to-end neural network architecture, which involves a multitude of hyperparameters. More advanced hyperparameter selection techniques such as Bayesian optimisation could therefore be explored [39] during the training. Graphical convolutional neural work (GNN) [40], which has been an emerging model in survival analysis, could also be explored in the future. In addition, it could be observed that DFS and OS share some common significant prognostic factors. This suggest the possibility of multi-task learning of these two prediction tasks [41]. Furthermore, introducing additional calibration techniques [37] could improve the alignment of predictions with ground truth data. Methods like SHAP [38] could also enhance the interpretability of neural networks by identifying crucial features in predictive models.

Acknowledgement. This work is funded by Cancer Research UK (CRUK). The authors acknowledge the contributions of Sinead King, St. James's Hospital, Dublin, Ireland; Hannah Adams, Gloucestershire Hospitals NHS Foundation Trust, England; and Masaru Hayami, CLINTEC, Karolinska Institutet, Stockholm, Sweden. The authors acknowledge the contributions to and previous works on the ENSURE study by Elliott J.A. et al. [16] and Gujjuri R.R. et al. [12]. The authors would like to thank the Oxford Biomedical Research Computing (BMRC) facility for providing the computing resources. Special thanks to Lav Radosavljevic from University of Oxford for his professional advice on statistical analysis.

Disclosure of Interests. The authors have no competing interests to declare that are relevant to the content of this article.

A Appendix

A.1 Imputation Procedure

Figure 1 illustrates the overall process of variable preprocessing and imputation, as described in Sect. 2.4.

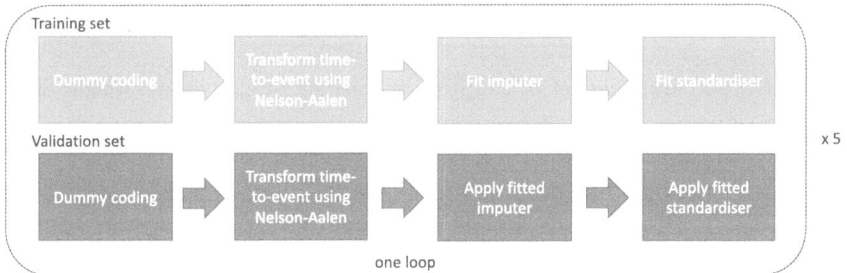

Fig. 1. Flowchart illustrating the preprocessing and imputation process. The process is performed for each loop and repeated across all loops within the CV.

A.2 Splitting Strategy

Figure 2 illustrates the splitting strategy during training for the two tasks.

A.3 Hyperparameter Tuning

Table 4 summarises the optimal set of hyperparameters of the three models. For DeepSurv and DeepHit, various network structures were explored, along with different number of epochs, batch sizes, optimiser schedulers, and learning rates.

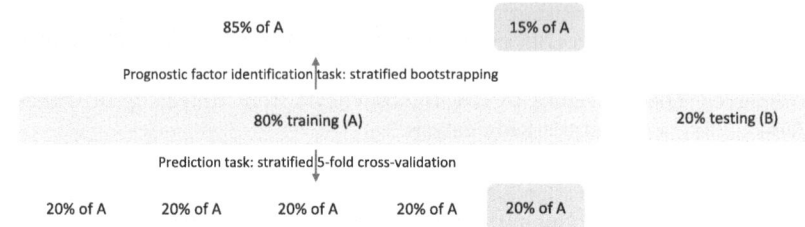

Fig. 2. Graphical illustration of the splitting strategy during training for prognostic factor identification task and prediction task, where blue and orange color represents the training and validation set for each task, respectively. (Color figure online)

Table 4. Summary of the optimal hyperparameter set of the three models for DFS and OS.

(a) Best hyperparameter set of CoxPH.

	L1 penalty	L2 penalty
DFS	0.008	0.001
OS	0.006	0.002

(b) Best hyperparameter set of DeepHit. lr: learning rate; w decay: weight decay.

	network	dropout	epochs	batch size
DFS	[64, 64]	0.1	75	64
OS	[64, 128, 64]	0.1	70	256
	optimiser	initial lr	scheduler	w decay
DFS	Adam	0.1	Exp.LR, $\gamma=0.7$	0.05
OS	Adam	0.1	Exp.LR, $\gamma=0.7$	0.05

(c) Best hyperparameter set of DeepHit. lr: learning rate; w decay: weight decay.

	network	dropout	epochs	batch size	optimiser
DFS	[64, 128, 64]	0.1	25	64	Adam
OS	[64, 128, 64]	0.1	100	64	Adam
	initial lr	scheduler	w decay	no. of output	output interp. no.
DFS	0.005	Exp.LR, $\gamma=0.7$	0.05	60	50
OS	0.005	Exp.LR, $\gamma=0.7$	0.05	60	50

A.4 Predicted Survival Curves

Figure 3 shows the predicted survival curves by three models of the same five patients.

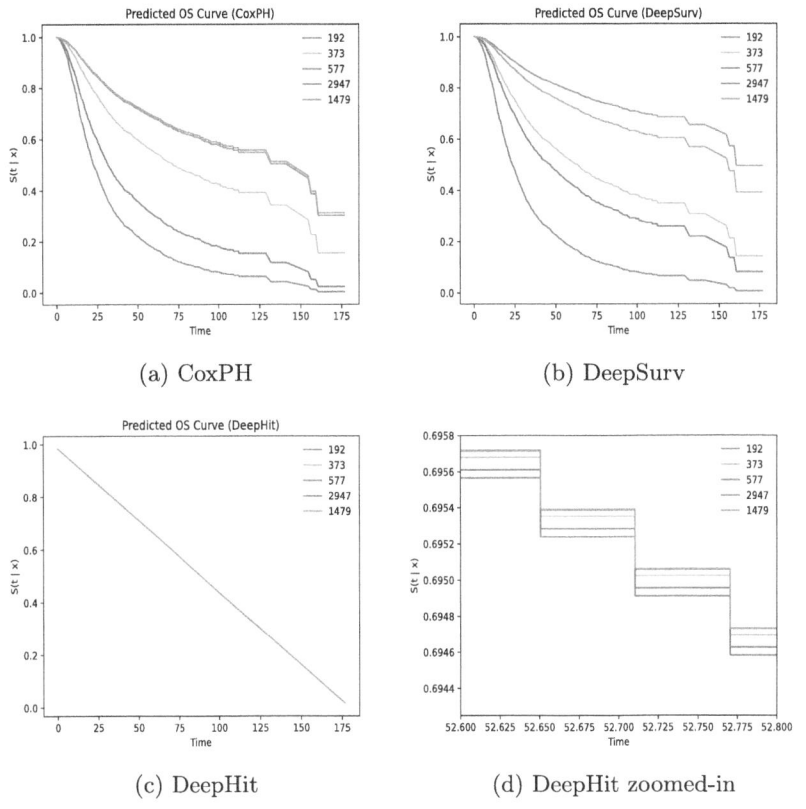

Fig. 3. Predicted OS curves for the same random set of five patients by three models respectively. The legend shows the patient ID.

References

1. Siegel, R., Giaquinto, A.N., Jemal, A.: Cancer statistics, 2024. CA Cancer J. Clin. **74**(1), 12–49 (2024)
2. Mariette, C., et al.: Factors predictive of complete resection of operable esophageal cancer: review of 746 patients. Gastroenterol. Clin. Biol. **26**(5), 454–462 (2002)
3. Boerner, T., et al.: Incidence and management of esophageal cancer recurrence to regional lymph nodes after curative esophagectomy. Int. J. Cancer **152**(10), 2109–2122 (2023)
4. Kunisaki, C., et al.: Surgical outcomes in esophageal cancer patients with tumor recurrence after curative esophagectomy. J. Gastrointest. Surg. **12**(5), 802–10 (2008)
5. Barbar, L., et al.: Prognostic immune markers for recurrence and survival in locally advanced esophageal adenocarcinoma. Oncotarget **10**(44), 4546–4555 (2019)
6. Wang, Q., Lagegren, J., Xie, S.: Prediction of individuals at high absolute risk of esophageal squamous cell carcinoma. Gastrointest. Endosc. **89**(4), 726–732 (2019)

7. Chen, W., et al.: Selection of high-risk individuals for esophageal cancer screening: a prediction model of esophageal squamous cell carcinoma based on a multicenter screening cohort in rural China. Int. J. Cancer **148**(2), 329–339 (2021)
8. Cox, D.R.: Regression models and life-tables. J. Roy. Stat. Soc.: Ser. B (Methodol.) **34**(2), 187–202 (1972)
9. Wang, W., et al.: A novel molecular and clinical staging model to predict survival for patients with esophageal squamous cell carcinoma. Oncotarget **7**(39), 63526–63536 (2016)
10. Gabriel, E., et al.: Novel calculator to estimate overall survival benefit from neoadjuvant chemoradiation in patients with esophageal adenocarcinoma. J. Am. Coll. Surg. **224**(5), 884–894 (2017)
11. Shapiro, J., et al.: Prediction of survival in patients with oesophageal or junctional cancer receiving neoadjuvant chemoradiotherapy and surgery. Br. J. Surg. **103**(8), 1039–47 (2016)
12. Gujjuri, R.R.: Predicting long-term survival and time-to-recurrence after esophagectomy in patients with esophageal cancer - Development and validation of a multivariate prediction model. Ann. Surg. **13**, 971–978 (2023)
13. Zhang, K., et al.: Machine learning-based prediction of survival prognosis in esophageal squamous cell carcinoma. Sci. Rep. **13**, 13532 (2023)
14. Gong, X., et al.: Application of machine learning approaches to predict the 5-year survival status of patients with esophageal cancer. JTD. **3**(11), 6240–6251 (2021)
15. Chen, T., Guestrin, C.: XGBoost: a scalable tree boosting system. In: Proceedings of the 22nd ACM SIGKDD International Conference on Knowledge Discovery and Data Mining, pp. 785–794 (2016)
16. Elliott, J.A., et al.: An international multicenter study exploring whether surveillance after esophageal cancer surgery impacts oncological and quality of life outcomes (ENSURE). Ann. Surg. **277**(5), 1035–1044 (2023)
17. Rice, T.W., Patil, D.T., Blackstone, E.H.: 8th edition AJCC/UICC staging of cancers of the esophagus and esophagogastric junction: application to clinical practice. Ann. Cardiothorac. Surg. **6**(2), 119–130 (2017)
18. Si, G, et al.: Progression-free survival: what does it mean for psychological well-being or quality of life? Agency for healthcare research and quality (US) (2013)
19. Lebwohl, D., et al.: Progression-free survival: gaining on overall survival as a gold standard and accelerating drug development. Cancer J. **15**(5), 386–94 (2009)
20. Dong, Y., Peng, C.J.: Principled missing data methods for researchers. SpringerPlus. **222**(2) (2013). https://doi.org/10.1186/2193-1801-2-222
21. Mack, C., Su, Z., Westreich, D.: Managing Missing Data in Patient Registries: Addendum to Registries for Evaluating Patient Outcomes: A User's Guide, Third Edition [Internet]. Rockville (MD): Agency for Healthcare Research and Quality (US) **17**(18), EHC015-EF (2018)
22. García-Laencina, P.J., et al.: Missing data imputation on the 5-year survival prediction of breast cancer patients with unknown discrete values. Comput. Biol. Med. **59**, 125–133 (2015)
23. Jerez, J.M., et al.: Missing data imputation using statistical and machine learning methods in a real breast cancer problem. Artif. Intell. Med. **50**(2), 105–115 (2010)
24. Azur, M.J., et al.: Multiple imputation by chained equations: what is it and how does it work? NT. J. Methods Psychiatr. **20**(1), 40–49 (2011)
25. Li, P., Stuart, E.A., Allison, D.B.: Multiple imputation: a flexible tool for handling missing data. JAMA **314**(18), 1966–7 (2015)
26. Rubin, D.B.: Flexible Imputation of Missing Data, 2nd edn. Chapman and Hall/CRC (2018). Multiple imputation

27. Austin, P.C., et al.: Missing data in clinical research: a tutorial on multiple imputation. Can. J. Cardiol. **37**(9), 1322–1331 (2021)
28. Colosimo, E., et al.: Empirical comparisons between Kaplan-Meier and Nelson-Aalen survival function estimators. J. Statist. Comput. Simul. **72**(4), 299–308 (2002)
29. White, I.R., Royston, P.: Imputing missing covariate values for the Cox model. Statist. Med. **28**(15), 1982–1998 (2009)
30. Spooner, A., et al.: A comparison of machine learning methods for survival analysis of high-dimensional clinical data for dementia prediction. Sci Rep. **10**, 20410 (2020)
31. Ake, C.F., et al.: Rounding after multiple imputation with non-binary categorical covariates, pp. 112–30 (2005)
32. Akoglu, H.: User's guide to correlation coefficients. Turkish J. Emerg. Med. **18**(3), 91–93 (2018)
33. Katzman, J.L., et al.: DeepSurv: personalized treatment recommender system using a Cox proportional hazards deep neural network. BMC Med. Res. Methodol. **18**(24), (2018). https://doi.org/10.1186/s12874-018-0482-1
34. Lee, C., et al.: DeepHit: a deep learning approach to survival analysis with competing risks. In: Proceedings of the AAAI Conference on Artificial Intelligence, pp. 2314–2321 (2018)
35. Smyth, E.C., et al.: Oesophageal cancer. Nat. Rev. Dis. Primers **3**, 1–21 (2017)
36. Zou, H.: Hastie, T.: Regularization and variable selection via the elastic net. J. R. Statist. Soc. B **67**(2), 301–320 (2005)
37. Goldstein, M., et al.: X-CAL: explicit calibration for survival analysis. Adv. Neural. Inf. Process. Syst. **67**(2), 18296–18307 (2020)
38. Lundberg, S.M.,Lee, S.: A unified approach to interpreting model predictions. In: Proceedings of the 31st International Conference on Neural Information Processing Systems, pp. 4768–4777. New York, USA (2017)
39. Kaur, P., Singh, A., Chana, I.: BSense: A parallel Bayesian hyperparameter optimized Stacked ensemble model for breast cancer survival prediction. Journal of Computational Science **60**,(2022)
40. Hou, W., et al.: Hybrid graph convolutional network with online masked autoencoder for robust multimodal cancer survival prediction. IEEE Trans. Med. Imaging **42**(8), 2462–2473 (2023)
41. Yun, S., Du, B., Mao, Y.: Robust Deep Multi-task Learning Framework for Cancer Survival Analysis. In: 2021 International Joint Conference on Neural Networks (IJCNN), pp. 1–8. Shenzhen, China (2021)
42. Furukawa, M., et al.: Prediction of recurrence free survival of head and neck cancer using PET/CT radiomics and clinical information (2024). https://arxiv.org/abs/2402.18417

Treatment Efficacy Prediction of Focused Ultrasound Therapies Using Multi-parametric Magnetic Resonance Imaging

Amanpreet Singh[✉], Samuel Adams-Tew, Sara Johnson, Henrik Odeen, Jill Shea, Audrey Johnson, Lorena Day, Alissa Pessin, Allison Payne, and Sarang Joshi

University of Utah, Utah, USA
{amanpreet.singh,samuel.adams,sara.l.johnson,henrik.odeen,jill.shea,
a.l.johnson,u1248745,u1471840,allison.payne,sarang.joshi}@utah.edu

Abstract. Magnetic resonance guided focused ultrasound (MRgFUS) is one of the most attractive emerging minimally invasive procedures for breast cancer, which induces localized hyperthermia, resulting in tumor cell death. Accurately assessing the post-ablation viability of all treated tumor tissue and surrounding margins immediately after MRgFUS thermal therapy residual tumor tissue is essential for evaluating treatment efficacy. While both thermal and vascular MRI-derived biomarkers are currently used to assess treatment efficacy, currently, no adequately accurate methods exist for the in vivo determination of tissue viability during treatment. The non-perfused volume (NPV) acquired three or more days following MRgFUS thermal ablation treatment is most correlated with the gold standard of histology. However, its delayed timing impedes real-time guidance for the treating clinician during the procedure. We present a robust deep-learning framework that leverages multiparametric MR imaging acquired during treatment to predict treatment efficacy. The network uses qualtitative T1, T2 weighted images and MR temperature image derived metrics to predict the three day post-ablation NPV. To validate the proposed approach, an ablation study was conducted on a dataset (N=6) of VX2 tumor model rabbits that had undergone MRgFUS ablation. Using a deep learning framework, we evaluated which of the acquired MRI inputs were most predictive of treatment efficacy as compared to the expert radiologist annotated 3 day post-treatment images.

Keywords: Focused Ultrasound · Treatment Efficacy · Computer-Aided Treatment Response

1 Introduction

Magnetic resonance guided focused ultrasound (MRgFUS) is a promising non-invasive ablation treatment for cancer. MRgFUS leverages magnetic resonance imaging (MRI) to plan, monitor and assess treatments under MRI guidance to attain tumor destruction with focused ultrasound. Because MRgFUS is applied non-invasively and the treated tumor is not resected, imaging biomarkers must determine tumor non-viability to ensure efficacious treatment. Imaging metrics currently used clinically in MRgFUS therapies include real-time MR thermometry, thermal dose, and T2-weighted and T1-weighted (with and without gadolinium contrast) images. However, numerous studies have shown these metrics, individually or combined, lack the necessary precision to ensure treatment efficacy, particularly when applied during the procedure. The literature presents conflicting changes in qualitative and quantitative MR metrics after MRgFUS ablation treatments, clearly documenting that this problem has not been adequately addressed.

Most ablative MRgFUS therapies are evaluated with imaging biomarkers derived from temperature [2,9] or vascular effects [12] of the treatment. Thermal dose is assessed cumulatively using the MR temperature images obtained during the ablation treatment [5,14], while vascular effects are typically assessed at the end of the treatment with T1-weighted contrast-enhanced (CE) MRI. Thermal dose has been shown to under-estimate the extent of tissue perfusion loss and necrosis in large treatment volumes [18], while CE-MRI in assessing immediate post-treatment tissue perfusion (<1 hr), is confounded by transient physiological responses including tissue edema, hemorrhage and cell membrane disruption [4]. While other MR contrasts are sensitive to acute physiological changes from MRgFUS treatment, measured changes have exhibited inconsistencies in the literature [3,4,15]. This evident sensitivity of multiple MR parameters to MRgFUS thermal ablation indicates that a non-contrast agent multiparametric imaging biomarker for immediately predicting tissue viability after MRgFUS is likely feasible.

Deep learning methods are revolutionizing several aspects of MRI [1,10,17] and this work is investigating the potential of a deep learning neural network imaging biomarker to predict MRgFUS ablation treatment outcomes. We present results demonstrating which components of the multiparametric imaging biomarker most accurately predict tissue viability. This information is critical as it will allow treatment providers the ability to streamline data acquisition during MRgFUS ablative therapies, improving the clinical workflow and efficiency of the procedure.

2 Methods

2.1 MRI Acquisition

All experiments were carried out with Institutional Animal Care and Use Committee approval. A New Zealand white rabbit VX2 tumor model was utilized (N=6). VX2 tumor cells were injected in the animals quadriceps muscle. After tumors had grown to 2 cm in length, they were ablated with a MRgFUS system inside a 3T MRI scanner (PrismaFIT Siemens, Erlangen, Germany). MRI guidance was utilized for tumor targeting, treatment monitoring, and post-treatment assessment. All MR sequences utilized are listed in Table 1. The rabbit was positioned on the treatment table as shown in Fig. 1. High-resolution (0.5 mm isotropic) non-contrast T1-weighted images of the entire treated leg were acquired prior to and immediately after ablation for image registration purposes. Coronal multi-parametric MR images of the lower leg (as seen in Fig. 1) were acquired immediately before, during, and 20 min after the ablation procedure. Approximately 40 min following the final sonication, CE-T1w MR images were acquired immediately following intravenous gadolinium injection (0.3 mL/kg ProHance) and animals were recovered. Due to the expected transient effects of ablation such as edema and latent apoptosis, which occurs for up to 72 hrs [7,11] after treatment, we defined the final treatment effect as the necrotic volume at 3 d after treatment. At this post-ablation time-point, the animal was re-positioned on the MRgFUS table and high resolution T1w and CE-T1w images were re-acquired.

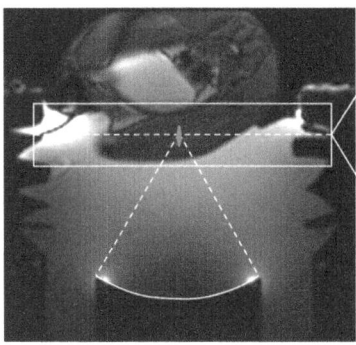
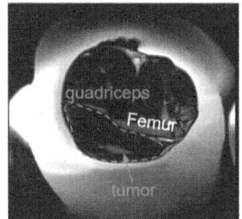

Fig. 1. MRgFUS thermal ablation setup. (left) Axial T2-weighted MRI localizer image showing the rabbit VX2 model positioned over the focused ultrasound transducer. The white lines indicate the acoustic beam path of the focused ultrasound, focusing to the orange point. The yellow box shows the approximate coronal acquisition volume used for all imaging inputs for the neural networks. (right) A coronal center slice of the T2-weighted pre-treatment image used for neural network training. The quadriceps, femur and tumor are all labeled. (Color figure online)

2.2 Non-perfused Volume Label

Semi-automatic segmentation of the non-perfused volume (NPV) – specifically the hypointense region at the treatment site – was assessed in the contrast-enhanced T1w image both in the ablation day and the 3 day time point. This was achieved by thresholding the CE-T1w image intensity to encompass the non-enhancing region, followed by manual editing. The 3 day time point NPV segmentations were considered the clinical NPV prediction of tissue necrosis and was used as the neural network training label, as described below. We defined the final treatment effect by the 3 day NPV volume.

2.3 Image Registration

A registration between all acquired data timepoints is necessary to create a dataset to train neural networks to predict treatment efficacy. The registration methods used in this work have been previously developed and validated. Briefly, between treatment day and 3 day follow-up imaging, there were unavoidable changes in the animal positioning in the MRI scanner. A volume-conserving longitudinal registration pipeline incorporating landmark-driven rigid registration, landmark-driven gradient flow [19] and image intensity-drive gradient flow [6], was used for registration, since tissue volume is preserved under normal physiological loading. The rigid registration is performed using 4–6 landmarks near the ends of the femur. The landmark-driven gradient flow utilizes the splines kernel which acts as a smoothing operator. The landmarks chosen for this step are blood vessel bifurcations and nutrient foramina, small openings in the bone through which blood vessels, nerves, and other structures access the bone marrow.

Table 1. Description of MR sequences used in the study. The MR temperature imaging is acquired as a multi-acquisition dynamic sequence during the entire length of each MRgFUS sonication. T1-weighted images were acquired both without and with gadolinium contrast.

MR image type	Sequence	Field of View (mm)	Resolution (mm)	Acq. Time (mm:ss)
T1-weighted (T1w)	3D VIBE	256 × 192 × 52	0.5 × 0.5 × 0.5	01:48
T2-weighted (T2w)	3D SPACE	256 × 192 × 52	1 × 1 × 1	02:32
MRI Temperature	3D GRE-EPI	192 × 150 × 20	2 × 1 × 3	00:04.5

To register the day 3 NPV volume to the ablation day data, the source image was chosen as the high-resolution T1w image without contrast, and the target image was a similar image acquired immediately after the MRgFUS ablation procedure. Registration was performed with the non-contrast images as these images do not include intensity features of the tumor or ablation sites, making the intensity-based registration unbiased to these features. The registration provided an invertible 3D diffeomorphism between the treatment day and 3 day follow up images. Validation was performed with a set of expert radiologist-selected landmarks in the target and source images, respectively. Ten source and target landmark pairs were chosen for each validation subject and the target registration error was calculated as the Euclidean distance between the deformed source point and the corresponding target landmark. Target registration error, evaluated over 8 animal subjects (separate from animals presented in this study as these subjects did not have all of the MRI imaging sequences acquired and could not be used for this study), was 0.93 ± 0.13 mm.

2.4 Neural Network Architecture: U-net

U-Net [13] segmentation architecture was used to generate the treatment efficacy biomarker using a combination of seven input channels: T1-weighted pre-

and post-treatment, T2-weighted pre- and post-treatment, gadolinium contrast enhanced T1-weighted post-treatment, Maximum Temperature Projection(MTP) and Cumulative Thermal Dose, both calculated from MR thermometry [14]. We used a 2 layer deep U-Net architecture for our experiments, with the convolutional layers having a kernel of size 3 and stride 1.

The network returns a 3D voxel-wise probability map of the tissue viability, and the output of the network was evaluated against the 3 day NPV label using a binary cross entropy loss function. To assess the generalizability of the U-Net biomarker, validation was performed using the Leave One Out (LOO) method. In each run, the training involved five rabbit subjects, while the evaluation was conducted on the remaining subject. During the evaluation, the network's probability output was thresholded at 0.5 to distinguish between viable and non-viable tissue. The results from different runs were aggregated to provide an overall evaluation of the network's performance, which are presented in Sect. 3.

2.5 Implementation Details

The deep learning framework was implemented with PyTorch 1.12.1 on an Ubuntu machine with NVIDIA A6000 GPU. The network trained from scratch using Binary Cross Entropy Loss and Adam Optimizer [8] with a learning rate of 0.001. The training was done for at most 400 epochs. The data-set and code used to train the U-Net Biomarker along with the trained models will be made publicly available here.

Table 2. List of combinations of MR Sequences used in biomarker experiments.

MR Sequence	UNET						
	1	2	3	4	5	6	7
T1-weighted pre treatment	✓	✓	×	✓	✓	×	✓
T1-weighted post treatment	✓	✓	×	✓	✓	×	✓
T2-weighted pre treatment	✓	✓	✓	×	×	✓	✓
T2-weighted post treatment	✓	✓	✓	×	×	✓	✓
CE-T1-weighted	✓	×	✓	✓	×	×	✓
Maximum Temperature Projection	×	✓	✓	✓	✓	✓	✓
Cumulative Thermal Dose	×	✓	✓	✓	✓	✓	✓

2.6 Biomarker Experiments

We ran an extensive set of experiments to find which MR images contained the most important information when it comes to predicting the treatment efficacy, through comparison to the 3 day NPV segmentation. A list of combinations of MR images used to inform the deep learning network is presented in Table 2. Each of these combinations was used as an input to train a U-Net in a supervised fashion against the 3 day NPV. We compared the predictive power of each of our experiments to the currently used thermal and vascular biomarkers using the Dice, Intersection over Union and the Mean Distance to agreement scores.

2.7 Evaulation Metrics and Data Analysis

Intersection over Union (IoU) is the ratio of the intersection of the two segmented areas to their combined areas. An IoU score of 1 indicates a perfect match between the segmentations, whereas a score of 0 means no overlap between the two.

Dice Score is a similarity index used for discrete data. It quantifies how well the predicted region aligns with the ground truth region. A Dice score of 1 indicates a perfect match between the segmentations, whereas a score of 0 means no overlap between the two.

Mean Distance to Agreement(MDA) is calculated by comparing the boundaries of the segments in two different segmentations [16]. It measures how close or distant the boundaries of corresponding segments are in two labels. The smaller the MDA, the higher the agreement between the segmentations.

Performance of the deep-learning based biomarkers was compared to the current clinically used thermal and vascular imaging biomarkers, specifically, the treatment day segmented NPV volume and the cumulative thermal dose (CTD), thresholded at 240 cumulative equivalent minutes at 43°C, using the metrics explained above. Comparison to both the currently used metrics to predict treatment efficacy is presented in Fig. 2.

3 Results

The results of each of the experiment U-Net combinations, compared to the thermal biomarker of cumulative thermal dose at 240 cumulative equivalent minutes and NPV computed immediately after the ablation is reported in Fig. 2. The U-Net (U-Net 2) that included pre- and post-treatment T1-weighted and T2-weighted data, as well as the maximum temperature predication and cumulative thermal dose yielded the best results. The mean Dice and IOU scores of 0.684 and 0.488 performed comparably to currently used clinical metrics of NPV (0.675 and 0.518 respectively). The lowest performing combination omitted any MR thermometry derived metrics (U-Net 1). These results are also seen as a function of MDA in Fig. 2. Figure 3 shows the spatial prediction in a single slice in one animal subject. The preidction is overlaid on various MR images used as inputs to the deep-learning based biomarker.

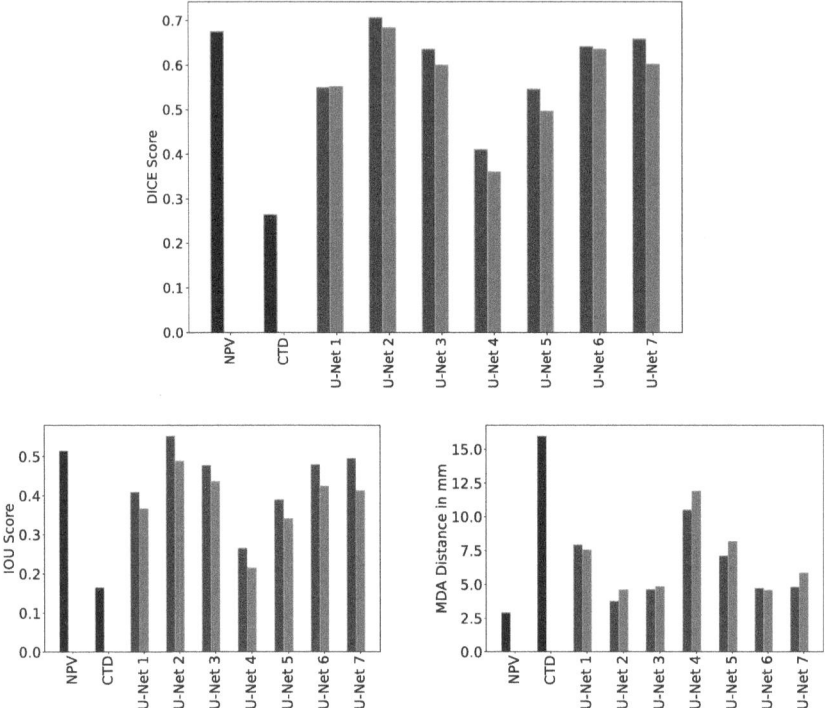

Fig. 2. Performance metrics of the LOO testing in 5 fold cross validations per experiment. The mean of both the training (blue) and testing (red) data sets are shown for each experiment. The performance of the NPV and CTD metrics (black) are shown as a reference. Dice (top), Intersection of Union (IoU, bottom left), and Mean Distance Agreement (MDA, bottom right) are shown. (Color figure online)

4 Discussion

The best results (U-Net 2) included all MR images except the contrast-enhanced T1-weighted image acquired immediately after the ablation procedure. This is notable, as the omitted contrast is often used to assess treatment efficacy. These results confirm the results of others that report acutely acquired contrast-enhanced T1-weighted images cannot be a reliable source of treatment effectiveness. It is also of note that U-Net 1, which did not include MR thermometry derived data, was the worst performer. It is clearly critical to include the subject-specific temperature increases in the treatment efficacy biomarker prediction.

Another interesting observation is between U-Net 6 and U-Net 7, which only differ in using the T1-weighted images for the 3 day NPV prediction. Addition of the T1-weighted data seems to have an adverse effect when used in conjunction with the contrast enhanced T1-weighted image. Therefore, care must be taken on which T1-weighted data should be used as an input for the biomarker prediction.

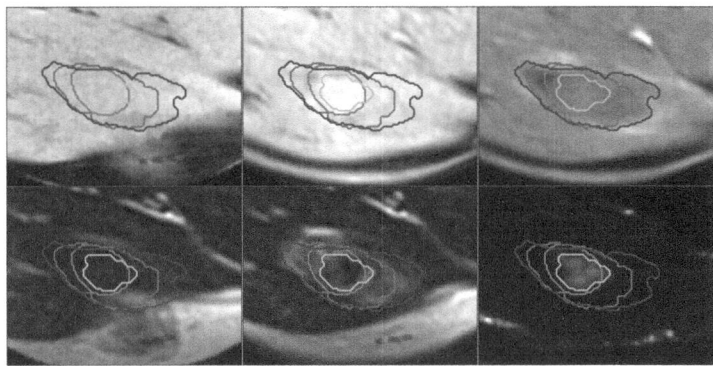

Fig. 3. Prediction example of U-Net-7 overlaid on Subject 4 MRI data. The green outline is the ground truth 3 day (Color figure online) NPV, the blue outline is the ablation day NPV, the yellow outline is the CTD biomarker while red outline is the output probability map of the network thresholded at 0.5. The sequences shown on top row left to right are: pre-treatment T1-weighted, post-treatment T1-weighted, contrast enhanced T1-weighted; bottom row: pre-treatment T2-weighted, post- treatment T2-weighted and MTP.

There are several clear study limitations, including the lack of quantitative MRI data, including T1 and T2 maps, and diffusion MRI. This study also presents a limited data set (N=6). While collecting data from more animal subjects will help further validate this presented biomarker, the limited number of data subjects is representative of longitudinal studies tracking patient outcomes. Longitudinal studies are often limited by patient follow-up adherence and the high costs associated with long-term monitoring. Similarly, while the VX2 rabbit tumor model exhibits rapid growth similar to certain types of human breast cancer, making it an effective model for studying tumor response to MRgFUS treatment, the high costs, need for specialized equipment and expertise to perform MRgFUS procedures on animals limits data acquisition. In addition, the variability in animal handling and care can introduce inconsistencies in the data, complicating the interpretation of results.

5 Conclusion and Future Work

The motivation behind this work was to investigate if multi-parametric MRI biomarkers could improve upon currently clinically utilized thermal and vascular biomarkers used to predict treatment efficacy in MRgFUS thermal ablation treatments. A volume preserving registration facilitated supervised machine learning to fully leverage the volumetric information available in multi-parametric MRI imaging across multiple time points. A U-Net style neural network using pre- and post-treatment T1w, T2w, and MR thermometry as inputs performs similarly to the currently used clinical metrics to evaluate tissue efficacy. Our results demonstrate use of MR-thermometry and T2w images are the most important in MRgFUS treatment efficacy prediction. Non-contrast MR biomarkers will allow more flexible non-invasive treatments, improving the accuracy of MRgFUS thermal ablation treatment of solid tumors.

For future work, we plan to expand our study to include a larger and more diverse dataset to enhance the robustness and generalizability of our deep-learning model. A key focus will be on comparing different prediction models to identify the most accurate and reliable approach for predicting treatment efficacy in MRgFUS. This comparison will involve evaluating various deep-learning architectures and incorporating additional MRI parameters (T1 and T2 maps, and diffusion MRI) that may improve prediction accuracy. Furthermore, we aim to use histology as the true label for assessing treatment efficacy, providing a direct comparison between our model predictions and the gold standard of tissue analysis. This will involve collecting histological data from post-treatment tissue samples and integrating it into our model validation process. By doing so, we aim to refine our framework and establish a comprehensive, real-time guidance system for clinicians during MRgFUS procedures, ultimately improving patient outcomes in breast cancer treatment.

Acknowledgement. Funded by the National Cancer Institute of the National Institutes of Health under Award Number R01CA259686.

Disclosure of Interests. The authors have no competing interests to declare that are relevant to the content of this article.

References

1. Chen, F., et al.: Variable-density single-shot fast spin-echo MRI with deep learning reconstruction by using variational networks. Radiology **289**(2), 366–373 (2018)
2. Eyerly, S.A., Vejdani-Jahromi, M., Dumont, D.M., Trahey, G.E., Wolf, P.D.: The evolution of tissue stiffness at radiofrequency ablation sites during lesion formation and in the peri-ablation period. J. Cardiovasc. Electrophysiol. **26**(9), 1009–1018 (2015)
3. Hazle, J.D., Stafford, R.J., Price, R.E.: Magnetic resonance imaging-guided focused ultrasound thermal therapy in experimental animal models: correlation of ablation volumes with pathology in rabbit muscle and VX2 tumors . J. Magn. Reson. Imaging Official J. Int. Soc. Magn. Reson. Med. **15**(2), 185–194 (2002)
4. Hectors, S.J., Jacobs, I., Moonen, C.T., Strijkers, G.J., Nicolay, K.: MRI methods for the evaluation of high intensity focused ultrasound tumor treatment: current status and future needs. Magn. Reson. Med. **75**(1), 302–317 (2016)
5. Henle, K., Dethlefsen, L.: Time-temperature relationships for heat-induced killing of mammalian cells. Ann. N. Y. Acad. Sci. **335**(1), 234–253 (1980)
6. Hinkle, J., Joshi, S.: Idiff: irrotational diffeomorphisms for computational anatomy. In: Information Processing in Medical Imaging: 23rd International Conference, IPMI 2013, Asilomar, CA, USA, June 28–July 3, 2013. Proceedings 23, pp. 754–765. Springer (2013). https://doi.org/10.1007/978-3-642-38868-2_63

7. Houssami, N., Macaskill, P., Luke Marinovich, M., Morrow, M.: The association of surgical margins and local recurrence in women with early-stage invasive breast cancer treated with breast-conserving therapy: a meta-analysis. Ann. Surg. Oncol. **21**, 717–730 (2014)
8. Kingma, D.P., Ba, J.: Adam: a method for stochastic optimization. arXiv preprint arXiv:1412.6980 (2014)
9. Lepock, J.R.: Cellular effects of hyperthermia: relevance to the minimum dose for thermal damage. Int. J. Hyperth. **19**(3), 252–266 (2003)
10. Litjens, G., et al.: A survey on deep learning in medical image analysis. Med. Image Anal. **42**, 60–88 (2017)
11. Lynn, J.G., Zwemer, R.L., Chick, A.J., Miller, A.E.: A new method for the generation and use of focused ultrasound in experimental biology. J. Gen. Physiol. **26**(2), 179 (1942)
12. McDannold, N.: Uterine leiomyomas: MR imaging-based thermometry and thermal dosimetry during focused ultrasound thermal ablation. Radiology **240**(1), 263–272 (2006)
13. Ronneberger, O., Fischer, P., Brox, T.: U-net: convolutional networks for biomedical image segmentation. In: Medical Image Computing and Computer-Assisted Intervention–MICCAI 2015: 18th International Conference, Munich, Germany, October 5-9, 2015, Proceedings, Part III 18, pp. 234–241. Springer (2015). https://doi.org/10.1007/978-3-319-24574-4_28
14. Sapareto, S.A., Dewey, W.C.: Thermal dose determination in cancer therapy. Int. J. Radiat. Oncol. Biol. Phys. **10**(6), 787–800 (1984)
15. Sequeiros, R.B., et al.: Liver tumor laser ablation-increase in the subacute ablation lesion volume detected with post procedural MRI. Acta Radiol. **51**(5), 505–511 (2010)
16. Taha, A.A., Hanbury, A.: Metrics for evaluating 3D medical image segmentation: analysis, selection, and tool. BMC Med. Imaging **15**(1), 1–28 (2015)
17. Yoo, S., Gujrathi, I., Haider, M.A., Khalvati, F.: Prostate cancer detection using deep convolutional neural networks. Sci. Rep. **9**(1), 19518 (2019)
18. Yoon, S.W., Cha, S.H., Ji, Y.G., Kim, H.C., Lee, M.H., Cho, J.H.: Magnetic resonance imaging-guided focused ultrasound surgery for symptomatic uterine fibroids: estimation of treatment efficacy using thermal dose calculations. Eur. J. Obstet. Gynecol. Reprod. Biol. **169**(2), 304–308 (2013)
19. Zimmerman, B.E., et al.: Learning multiparametric biomarkers for assessing MR-guided focused ultrasound treatment of malignant tumors. IEEE Trans. Biomed. Eng. **68**(5), 1737–1747 (2020)

SurRecNet: A Multi-task Model with Integrating MRI and Diagnostic Descriptions for Rectal Cancer Survival Analysis

Runqi Meng[1], Zonglin Liu[2,5], Yiqun Sun[1,2,5], Dengqiang Jia[3], Lin Teng[1], Qiong Ma[2,5], Tong Tong[2,5], Kaicong Sun[1], and Dinggang Shen[1,4(✉)]

[1] School of Biomedical Engineering and State Key Laboratory of Advanced Medical Materials and Devices, ShanghaiTech University, Shanghai, China
[2] Department of Radiology, Fudan University Shanghai Cancer Cente, Shanghai, China
[3] Hong Kong Centre for Cerebro-cardiovascular Health Engineering, Hong Kong, China
[4] Shanghai United Imaging Intelligence Co., Ltd., Shanghai, China
dinggang.shen@gmail.com
[5] Department of Oncology, Shanghai Medical College, Fudan University, Shanghai, China

Abstract. Survival analysis is paramount for cancer patients as it offers crucial prognostic insights for treatment planning. The performance of existing survival analysis methods is mainly limited by two factors: 1) inefficient extraction of features from multi-modal medical data, *e.g.*, MR images and clinical diagnostic descriptions; and 2) inadequate focus on disease-relevant regions, *e.g.*, primary tumor. To deal with these challenges, in this study, we propose a rectal cancer survival analysis model, dubbed as SurRecNet, which effectively fuse MR images and diagnostic descriptions and takes advantage of multi-task learning. Specifically, we introduce a cross-modality alignment module, aiming to precisely align diagnostic descriptions with MR images at a granular level and facilitate accurate survival analysis. Furthermore, SurRecNet simultaneously predicts tumor masks, relapse states, and survival outcomes by leveraging multi-task learning strategy, imitating the diagnostic process of radiologists to enhance prediction performance. Experimental results on a real clinical rectal multi-modal dataset demonstrate that our SurRecNet significantly outperforms the state-of-the-art methods.

Keywords: Survival Analysis · Vision-language Alignment · Rectal Cancer Analysis · Multi-task learning

---

R. Meng and Z. Liu—These authors contributed equally to this work.

© The Author(s), under exclusive license to Springer Nature Switzerland AG 2025
S. Ali et al. (Eds.): CaPTion 2024, LNCS 15199, pp. 200–210, 2025.
https://doi.org/10.1007/978-3-031-73376-5_19

1 Introduction

Rectal cancer is a prevalent malignancy within the gastrointestinal system, presenting a substantial risk of mortality [1]. Survival analysis plays an important role in the therapeutic process of rectal cancer, not only in guiding the formulation of personalized treatment strategies, but also in optimizing the clinical monitoring and dynamic management. [2]. In previous clinical practice, essential diagnostic information, including diagnostic descriptions extracted from radiology reports and clinical records, has had limited value as a prognostic tool due to its inability to fully reflect biological behavior. Magnetic resonance imaging is a non-invasion in vivo imaging, which can reflect the biological characteristics of rectal cancer, but it has not been well utilized in survival analysis. Therefore, given the complexity of clinical data on rectal cancer and limited specialized expertise, there emerges an exigent necessity to craft an automated and precise framework for the facilitation of accurate survival analysis.

Previous research has focused on radiomics analysis, which is a method of using data characterization algorithms to extract high-throughput features from imaging and establish statistical analysis models. Firstly, radiomics analysis entail extracting handcrafted features from pre-segmented tumor regions of MRI [3]. Then,these features are utilized for survival analysis, often em- ploying statistical techniques such as the Cox Proportional Hazard (CoxPH) [4] model. In recent years, the advent of deep learning has sparked a surge of innovation, leading to several pioneer works [5–8] for automatic rectal cancer survival prediction, which bypasses the need for pre-segmented tumor masks. Despite these advancements, there still appear to be significant limitations. First, many methods struggle to guide neural networks toward disease-relevant regions in imaging data without relying on tumor masks. Second, there is a lack of robust techniques for the comprehensive exploration of multi-modal medical data, including both imaging data and diagnostic descriptions. These limitations largely hinder the performance of deep learning-based methods for survival analysis.

In conventional clinical research, radiologists typically manually segment tumor areas and then assess the probability of relapse and survival risk using MR images in conjunction with tumor masks and other relevant medical data (*e.g.*, radiology reports, gender, age, etc.) [9]. While deep survival models offer the advantage of conducting end-to-end survival predictions without necessitating tumor masks, they face challenges in extracting tumor-specific details (*e.g.*, prognostic information from primary tumor regions), and thus lack adequate discriminative evidence for direct survival analysis [10]. To overcome this limitation, recent deep survival models have embraced multi-task learning [11–13], concurrently tackling tumor segmentation and survival prediction to guide feature extraction to focus on relevant tumor regions implicitly. However, these models directly fuse multi-modal features without explicit alignment, potentially limiting the performance of the models.

Recently, CLIP [21] introduced a strategy to measure the similarity between image and text features. This shows that natural language supervision effectively transfers linguistic information to visual representations [15], with significant implications for aligning medical images with diagnostic descriptions. Unlike natural images, medical images often present with restricted semantic regions, where background actually predominates and disease-related areas are typically small. To address these challenges, several methods [15–17] introduce a prototype representation learning framework that encompasses alignment between 2D X-ray images and their corresponding reports. However, aligning imaging and diagnostic descriptions in 3D space poses notable challenges, which demand a finer level of alignment granularity.

To tackle the aforementioned challenges, in this study, we introduce SurRecNet, a network devised for rectal cancer survival analysis, which integrates T2-weighted images (T2WIs) and diagnostic descriptions in a multi-task learning fashion. Specifically, our SurRecNet is built upon a multi-task learning framework, imitating the clinical diagnostic process and facilitating the explicit capture of complex multi-modal features in survival analysis. Furthermore, to establish fine-grained correspondence between MR images and diagnostic descriptions, we introduce a cross-modality representation alignment module to align diagnostic descriptions with imaging data at the imaging patch level and sentence level. Experimental evaluations on our in-house dataset demonstrate the effectiveness of the proposed approach in accurately predicting patients' survival outcomes, significantly outperforming alternative methods.

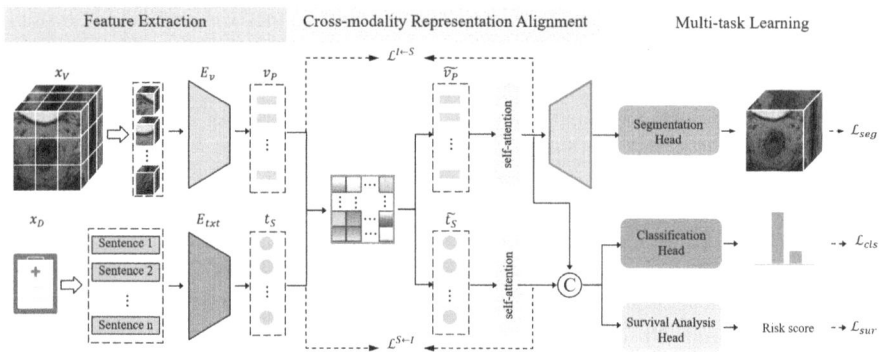

Fig. 1. An overview of the proposed SurRecNet. Our model exploits the incorporated information of MR images and clinical reports to perform tumor segmentation, relapse state classification, and survival prediction based on multi-task learning.

2 Method

The architecture of the proposed multi-task model consists of three main components including feature extraction, cross-modality representation alignment, and multi-task learning as illustrated in Fig. 1. Given the distinct representations between T2WIs and diagnostic descriptions, the feature extraction stage employs separate encoding branches for each modality, as detailed in Sect. 2.1. Subsequently, cross-modality representation alignment is conducted to locate image and diagnostic description features in the latent space, as outlined in Sect. 2.2.

Our model is trained to jointly perform tumor segmentation, relapse state classification, and survival prediction simultaneously with task-specific head branches, as described in Sect. 2.3. Section 2.4 describes the loss functions utilized in our work. More details on each part are provided in the following sections.

2.1 Feature Extraction

Given an image x_v and its corresponding clinical diagnostic descriptions x_d, both imagery and textual features are independently extracted using separate branches. Notably, in original T2WIs, disease-related regions tend to be extremely small-scale. To precisely align image and description features in the common space, we utilize 3D patches and sentences as input for the imaging encoder and text encoder, respectively.

Visual feature extraction. Given an image $x_V \in \mathbb{R}^{H \times W \times D \times C}$, where H, W, D, and C represent the height, width, depth, and channel number of the 3D image, respectively, we divide it into $m = HWD/p^3$ non-overlapping 3D patches $x_P = \{x_1, \cdots, x_m\}$, with $x_i \in \mathbb{R}^{p \times p \times p}$ being the patch. The visual encoder E_v encodes these patches into sub-regional representations $v_P = \{v_1, \cdots, v_m\}$. In this work, we adopt the encoder of UNETR [18] to extract image features.

Textual features extraction. With respect to clinical diagnostic descriptions x_D, sentences are tokenized and passed through a text encoder E_{txt}, to obtain sentence-wise textual features $t_S = \{t_1, \cdots, t_n\}$ at the token level. In our implementation, we use BioBERT [19] as our textual encoder to compute the embedding for each sentence.

2.2 Cross-Modality Representation Alignment

Given local representations from both imaging and textual modalities v_P and t_S, our objective is to align features of the clinical diagnostic descriptions and image regions. We achieve this alignment by performing localized cross-modality information interaction and maximizing the mutual information across patch-sentence pairs. Since not all sub-regions and sentences contribute equally to semantic understanding, we employ the Sigmoid function $\sigma(\cdot)$ instead of Softmax function inspired by [15].

We obtain the interacted features by cross-attention mechanisms. Specifically, the local sentence-to-patch and patch-to-sentence attention-based representations can be formulated as follows:

$$\widetilde{v_i} = \sum_{j=1}^{n} \sigma\left(\frac{W_{Q1}v_i \cdot W_{K1}t_j^T}{\sqrt{D}}\right) \cdot W_{V1}t_j \tag{1}$$

$$\widetilde{t_j} = \sum_{i=1}^{m} \sigma\left(\frac{W_{Q2}t_j \cdot W_{K2}v_i^T}{\sqrt{D}}\right) \cdot W_{V2}v_i \tag{2}$$

Here, W_{Q1} and W_{Q2} are query weight matrices for imaging and textual features, while W_{K1}, W_{K2}, W_{V1}, and W_{V2} denote key and value weight matrices for both feature types, respectively. These parameters are learned during training process.

Then, we apply a simple self-attention mechanism to the cross-modality representations in preparation for bidirectional alignment. Given that the cross-modality attention representation encompasses both imaging and diagnostic features, we conduct local alignment on these representations using contrastive loss, with the objective of maximizing the similarity between corresponding patch-sentence pairs and minimizing it between non-corresponding pairs. Specifically, for patch-to-sentence alignment, the cross-modality, and localized linguistic representations of the same sentence are regarded as a positive pair, while those from different sentences are negative pairs. Similarly, for sentence-to-patch alignment, matching cross-modality and localized imaging representations of the same patches are positive pairs, while those from different patches are negative pairs. The detailed losses are given below:

$$\mathcal{L}^{I \leftarrow S} = -\frac{1}{m} \sum_{i=1}^{m} \log \frac{\exp(v_i \cdot \widetilde{v_i}/\tau)}{\sum_{k=1}^{m} \exp(v_i \cdot \widetilde{v_k}/\tau)} \qquad (3)$$

$$\mathcal{L}^{S \leftarrow I} = -\frac{1}{n} \sum_{j=1}^{n} \log \frac{\exp(t_j \cdot \widetilde{t_j}/\tau)}{\sum_{k=1}^{n} \exp(t_j \cdot \widetilde{t_k}/\tau)} \qquad (4)$$

where τ is the temperature parameter. For local imaging representation v_i, the objective of $\mathcal{L}^{I \leftarrow S}$ is to enhance the similarity between v_i and its associated cross-modality attention representation, $\widetilde{v_i}$, while minimizing the similarity to other attention representations $\widetilde{v_{x \neq i}}$. The same principle applies to $\mathcal{L}^{S \leftarrow I}$.

Combing Eq (4) and Eq (3), the total contrastive loss is formulated as:

$$\mathcal{L}^{S||I} = \mathcal{L}^{I \leftarrow S} + \mathcal{L}^{S \leftarrow I} \qquad (5)$$

2.3 Multi-task Learning

In order to guide imaging feature extraction to focus on relevant tumor regions and also imitate the clinical diagnostic process, we employ multi-task learning strategy to perform tumor segmentation, relapse state classification, and survival analysis.

Specifically, we forward the aligned representations to a segmentation decoder of UNETR to emphasize tumor regions. In survival analysis tasks, it has been shown that discretizing time into a finite number of bins and treating it as classification tasks can lead to better performance compared to traditional continuous-time regression models [20]. Therefore, both the classification head and survival analysis head share the same structure, i.e., a 3D ResNet-based backbone network followed by a fully connected layer.

2.4 Loss Function

In the segmentation loss $\mathcal{L}_{seg}$, we combine Dice loss ($\mathcal{L}_{Dice}$) [24] and Focal loss ($\mathcal{L}_F$) [25]. For classification loss $\mathcal{L}_{cls}$, we employ the Focal loss [25]. Consistent with prior studies [11], we use the negative-log likelihood loss [26], denoted as $\mathcal{L}_{sur}$, for survival prediction. By combing all the loss functions described above, the final loss $\mathcal{L}_{overall}$ used for network training is expressed as follows:

$$\begin{aligned} \mathcal{L}_{overall} &= \mathcal{L}_{seg} + \mathcal{L}_{cls} + \mathcal{L}_{sur} + \lambda \mathcal{L}^{S||I} \\ \mathcal{L}_{seg} &= \mathcal{L}_{Dice} + \mathcal{L}_F \end{aligned} \quad (6)$$

3 Experiment

3.1 Dataset

Our dataset contains 373 cases, which are collected from three institutions. All cases are confirmed to have rectal tumors through histological biopsy analysis of surgically resected specimens, with accompanying diagnostic texts available. Tumor annotations are meticulously marked on the axial T2WIs by three experienced radiologists, drawing upon insights from biopsy pathology. These annotations undergo rigorous cross-validation among radiologists to establish a definitive and reliable ground truth.

To partition the dataset, we randomly allocate 261 cases for training, 37 cases for validation, and 75 cases for testing. We resampled the images into $0.3516 \times 0.3516 \times 3.3\,mm^3$ to have homogeneous spatial resolution and cropped out the background to have the image size of $256 \times 256 \times 19$. In contrast to recent works that solely align radiology reports with images, we construct diagnostic descriptions by pre-processing both radiology reports and medical information using a standardized clinical template. This approach ensures that any influence stemming from variations in radiologists' expression styles is minimized.

3.2 Evaluation Metrics

In this study, we use Dice Similarity Coefficient (DSC) and 95% Hausdorff Distance (HD95) for segmentation assessment, specificity and accuracy for classification, and Area Under the Curve of Receiver Operating Characteristic (AUC) and Concordance index (C-index) for survival prediction evaluation.

3.3 Implementation Details

All the experiments are implemented using PyTorch on two NVIDIA Tesla V100 GPUs equipped with 40GB RAM. We train the model for 500 epochs and use Adam optimizer with a learning rate of 5×10^{-5}. The batch size is set as 8 and the weighting parameter λ in the loss function is chosen as 0.95 based on experience. The temperature coefficients τ are empirically tuned as 0.01.

Table 1. Comparison of different methods, with the best outcomes highlighted in bold. As previous works rarely report results for all three tasks, we denote the missing results in their original work as 'N/A'.

Method	Survival Prediction		Classification		Segmentation	
	AUC [%]↑	C-index [%]↑	specificity [%]↑	accuracy [%]↑	DSC [%]↑	HD95 [mm]↓
CoxPH	72.28 ± 0.72	76.45 ± 0.52	N/A	N/A	N/A	N/A
DeepHit	70.56 ± 0.69	74.28 ± 0.41	N/A	N/A	N/A	N/A
OSPred	75.88 ± 0.82	77.44 ± 0.47	86.46 ± 0.32	80.46 ± 0.18	N/A	N/A
DeepMTS	73.55 ± 0.44	75.26 ± 0.59	N/A	N/A	68.19 ± 0.34	4.56 ± 0.64
TMSS	75.49 ± 0.63	75.24 ± 0.46	N/A	N/A	64.59 ± 0.30	3.22 ± 0.43
XSurv	75.19 ± 0.59	76.49 ± 0.52	N/A	N/A	71.03 ± 0.46	2.85 ± 0.33
Ours	**77.42 ± 0.54**	**79.02 ± 0.44**	**88.56 ± 0.21**	**83.52 ± 0.14**	**72.42 ± 0.43**	**2.54 ± 0.32**

3.4 Comparison with State-of-the-Art Methods

To evaluate the effectiveness of our model, we conduct comparative experiments with several state-of-the-art (SOTA) survival analysis methods, which can be roughly classified into two distinct categories: single-task and multi-task approaches. Specifically, we utilize CoxPH [22] and DeepHit [23] as representative single-task methods, while OSPred [14], DeepMTS [13], TMSS [11], and XSurv [12] serve as representative multi-task methods in our evaluation. To achieve a fair comparison of the original methods with single-modality input, we combine the diagnostic tabular data and radiomics features extracted from the MR images as the input.

We summarize the quantitative comparison in Table 1. We can see that our SurRecNet outperforms all the compared methods for all the tasks, which demonstrates that our SurRecNet can locate rectal tumor regions more precisely and has improved survival prediction. We attribute these performance improvements mainly to our proposed cross-modality representation alignment.

To figure out the attention regions of our network, we generate heatmaps for each case using Class Activation Mapping (CAM). As depicted in Fig. 2, four cases are presented, each displaying consecutive T2WI slices alongside their corresponding heatmaps derived from features of the self-attention layer. Detailed

Table 2. The detailed clinical information of the cases in Fig. 2, where AC represents postoperative adjuvant chemotherapy, RM represents recurrence or metastasis, and DFS represents disease-free survival.

Cases	Gender	Age	Clinical TN stage	Pathological TN stage	pTRG	AC	RM	DFS (months)
A	Male	45	cT4aN2	ypT3N1b	3	(+)	yes	8
B	Male	38	cT3N2	ypT3N2a	2	(+)	yes	20
C	Female	66	cT4aN2	ypT3N1a	2	(+)	no	44
D	Male	73	cT3aN1b	ypT3N1a	2	(+)	no	71

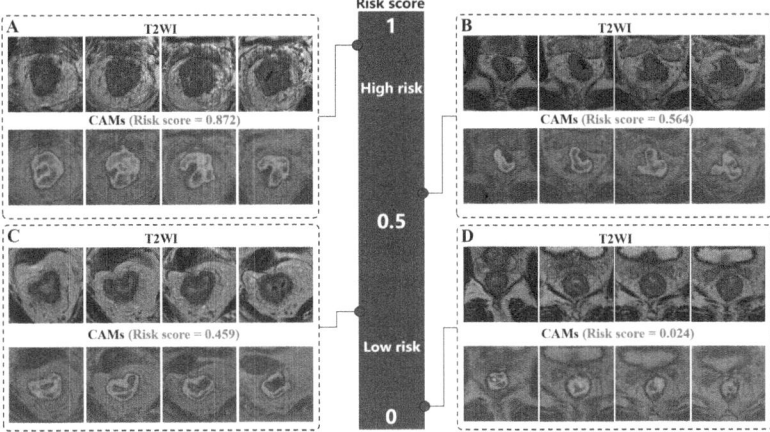

Fig. 2. Visual representation of four cases (from A to D) from the testing dataset. Each case exhibits four consecutive slices of T2WI (top row), along with their corresponding heatmaps (bottom row). The colored gradient bar indicates the risk score, with blue indicating a low risk and red indicating a high risk. The closer the risk score approaches to 1, the greater the risk of early occurrence of RM becomes. (Color figure online)

diagnostic clinical information for each patient is summarized in Table 2. It is evident that the heatmaps closely correspond to the tumor regions, indicating the effective extraction of features from disease-related areas by our network. Additionally, predicted risks align well with patients' DFS, with case A exhibiting a risk of up to 0.8 and a DFS of only 8 months, while case D shows a risk of 0.024 and a DFS of 71 months.

3.5 Ablation Study

To analyze the effectiveness of each components of our model, we conduct an ablation study for three model variants: 1) utilizing a single modality, solely T2WI (bNet-I) or diagnostic descriptions (bNet-D), 2) employing concatenated modalities (bNet-Full), and 3) combining two input modalities with cross-modality representation alignment (bNet-Full-CM) or multi-task learning (bNet-Full-MT). The quantitative comparisons of these various variants are listed in Table 3.

To assess the impact of imaging and text modalities, we utilize the corresponding encoder and a survival head for either image or text input. We can see that results using single-modality are notably poor. However, upon combining images and diagnostic sentences, a significant improvement is observed. Furthermore, the use of cross-modality representation alignment leads to a noteworthy improvement of 5.17% in AUC and 2.81% in C-index. Notably, a substantial improvement of 1.97% in AUC is attained with multi-task learning, indicating its efficacy in learning accurate imaging-text correspondences.

Table 3. Ablation study of key components. The best results are in **bold**.

Method	Modality		Survival Prediction	
	imaging	text	AUC [%]↑	C-index [%]↑
bNet-I	✓	✗	54.29 ± 0.56	58.32 ± 0.82
bNet-D	✗	✓	61.76 ± 0.72	65.32 ± 0.43
bNet-Full	✓	✓	70.28 ± 0.32	74.46 ± 0.29
bNet-Full-CM	✓	✓	75.45 ± 0.34	77.27 ± 0.46
bNet-Full-MT	✓	✓	74.49 ± 0.88	76.92 ± 0.76
bNet-Full-CM-MT	✓	✓	**77.42 ± 0.54**	**79.02 ± 0.44**

3.6 Conclusion

In this study, we propose a multi-task model for rectal cancer survival analysis, called SurRecNet, which combines MRI scans and diagnostic descriptions and is built on multi-task learning. Specifically, our approach incorporates MR images and diagnostic descriptions with a dedicated module for cross-modality representation alignment, which facilitates survival analysis. Additionally, our survival analysis model benefits from tumor segmentation and relapse state classification. We validate the effectiveness of our method through detailed analysis of four clinical cases. Experiments demonstrate the superiority of our method over other SOTA multi-modal approaches in survival analysis by effectively aligning and fusing multi-modal clinical data for rectal cancer. Although our method has shown promising performance for rectal cancer survival analysis, there are still some limitations. For example, the latent temporal relationships among patients remain underexplored in the survival analysis task. In the future, we may design a distance metric that can simultaneously identify significant features and measure differences in survival time among patients, to improve the precision of survival prediction.

Acknowledgments. This work was supported in part by National Natural Science Foundation of China (grant numbers 62250710165, U23A20295), and The Key R&D Program of Guangdong Province, China (numbers 2023B0303040001, 2021B0101420006).

Disclosure of Interests. The authors have no competing interests to declare that are relevant to the content of the article.

References

1. Hossain, M.D., et al.: Colorectal cancer: a review of carcinogenesis, global epidemiology, current challenges, risk factors, preventive and treatment strategies. Cancers **14**(7), 1732 (2022)
2. Gillies, R.J., et al.: Radiomics: images are more than pictures, they are data. Radiology **278**, 563–577 (2015)

3. Suter, Y., et al.: Radiomics for glioblastoma survival analysis in pre-operative MRI: exploring feature robustness, class boundaries, and machine learning techniques. Cancer Imaging **20**, 1–13 (2020)
4. Nadjib Bustan, M., et al.: Cox proportional hazard survival analysis to inpatient breast cancer cases. In: Journal of Physics: Conference Series, vol. 1028 (2018)
5. Suliman, W., Ravi, V., Luo, B., Sun, X.F., Pham, T.D.: Convolutional neural networks and support vector machines for five-year survival analysis of metastatic rectal cancer. In: 2022 International Joint Conference on Neural Networks (IJCNN), Padua, Italy, pp. 1–8, (2022). https://doi.org/10.1109/IJCNN55064.2022.9892935.
6. Pham, T.D.: Prediction of five-year survival rate for rectal cancer using markov models of convolutional features of RhoB expression on tissue microarray. In: IEEE/ACM Transactions on Computational Biology and Bioinformatics, vol. 20, no. 5, pp. 3195–3204 (2023). https://doi.org/10.1109/TCBB.2023.3274211.
7. Li, H.: Deep convolutional neural networks for imaging data based survival analysis of rectal cancer. In: IEEE 16th International Symposium on Biomedical Imaging (ISBI 2019), Venice, Italy, 846–849 (2019). https://doi.org/10.1109/ISBI.2019.8759301
8. Pham, T.D., Ravi, V., Fan, C., Luo, B., Sun, X.-F.: Classification of IHC Images of NATs With ResNet-FRP-LSTM for Predicting Survival Rates of Rectal Cancer Patients. IEEE J. Transl. Eng. Health Med. **11**, 87–95 (2023). https://doi.org/10.1109/JTEHM.2022.3229561
9. Liu, X., et al.: Deep learning radiomics-based prediction of distant metastasis in patients with locally advanced rectal cancer after neoadjuvant chemoradiotherapy: a multicentre study. EBioMedicine **69** (2021)
10. Deepa, P., Gunavathi, C.: A systematic review on machine learning and deep learning techniques in cancer survival prediction. Prog. Biophys. Mol. Biol. **174**, 62–71 (2022)
11. Saeed, N., et al.: TMSS: An End-to-End Transformer-Based Multimodal Network for Segmentation and Survival Prediction. ArXiv abs/2209.05036 (2022)
12. Meng, M., et al.: Merging-Diverging Hybrid Transformer Networks for Survival Prediction in Head and Neck Cancer. ArXiv abs/2307.03427 (2023)
13. Meng, M., et al.: DeepMTS: deep multi-task learning for survival prediction in patients with advanced nasopharyngeal carcinoma using pretreatment PET/CT. IEEE J. Biomed. Health Informatics **26**, 4497–4507 (2021)
14. Tang, Z., et al.: Pre-operative survival prediction of diffuse glioma patients with joint tumor subtyping. In: International Conference on Medical Image Computing and Computer-Assisted Intervention, pp. 786–795 (2023)
15. Cheng, P., et al.: Prior: prototype representation joint learning from medical images and reports. 2023 IEEE/CVF International Conference on Computer Vision (ICCV), pp. 21304–21314 (2023)
16. Wu, C., et al.: Medklip: medical knowledge enhanced language-image pre-training for X-ray diagnosis. In: 2023 IEEE/CVF International Conference on Computer Vision (ICCV), pp. 21315–21326 (2023)
17. Dawidowicz, G., et al.: Limitr: leveraging local information for medical image-text representation. In: 2023 IEEE/CVF International Conference on Computer Vision (ICCV), pp. 21108–21116 (2023)
18. Hatamizadeh, A., et al.: Unetr: transformers for 3D medical image segmentation. In: 2022 IEEE/CVF Winter Conference on Applications of Computer Vision (WACV), pp. 1748–1758 (2021)
19. Lee, J., et al.: BioBERT: a pre-trained biomedical language representation model for biomedical text mining. Bioinformatics **36**, 1234 –1240 (2019)

20. Suresh, K., et al.: Survival prediction models: an introduction to discrete-time modeling. BMC Med. Res. Methodol. **22** (2022)
21. Radford, A., et al.: Learning transferable visual models from natural language supervision. In: International Conference on Machine Learning, pp. 8748–8763 (2021)
22. Cox, D.R.: Regression models and life-tables. J. Roy. stat. Soc. Ser. B-Methodological **34**, 187–220 (1972)
23. Lee, C., et al.: Deephit: a deep learning approach to survival analysis with competing risks. In: AAAI Conference on Artificial Intelligence (2018)
24. Milletarì, F., et al.: V-net: fully convolutional neural networks for volumetric medical image segmentation. In: 2016 Fourth International Conference on 3D Vision (3DV), pp. 565-571 (2016)
25. Lin, T., et al.: Focal loss for dense object detection. In: 2017 IEEE International Conference on Computer Vision (ICCV), pp. 2999-3007 (2017)
26. Gensheimer, M.F., Narasimhan, B.: A scalable discrete-time survival model for neural networks. PeerJ **7**, e6257 (2018)

Improved Prediction of Recurrence After Prostate Cancer Radiotherapy Using Multimodal Data and *in Silico* simulations

Valentin Septiers[1,3](✉), Carlos Sosa-Marrero[1], Renaud de Crevoisier[1], Aurélien Briens[2], Hilda Chourak[1], Maria A. Zuluaga[3], and Oscar Acosta[1]

[1] Univ Rennes, CLCC Eugène Marquis, Inserm, LTSI - UMR 1099, 35000 Rennes, France
valentin.septiers@univ-rennes.fr
[2] CLCC Eugène Marquis, 35000 Rennes, France
[3] EURECOM, Data Science Department, 06410 Biot, France

Abstract. Prediction of biochemical recurrence (BCR) after prostate cancer radiotherapy is crucial for devising personalised treatments. BCR has been traditionally predicted using clinical data or *in vivo* imaging within AI frameworks such as radiomics approaches, but with limited results and reduced interpretability. These analysis are additionally hindered by the imbalanced and heterogeneous nature of data. In this paper, we present a novel approach to predict BCR at 5 years, based not only on clinical and image features, but also on a patient specific radiobiological mechanistic *in silico* model simulating tumour growth and radiation response. By combining all these data, we aim at i) improving the prediction of BCR after prostate cancer radiotherapy (RT), and ii) bringing interpretability to this prediction. A cohort of 254 patients was used. Pre-treatment T2-w MRIs, ADC maps and 7 clinicopathological characteristics were available. Patient specific digital twins of tumours were created from MRIs. The prescribed treatment was simulated with the mechanistic model yielding 414 features characterising the response of the tumour to RT. A first univariate feature selection analysis was conducted to select the most predictive features. Then, a machine learning algorithm was trained using selected features and compared with a deep learning (DL) approach based on clinicopathological characteristics and MRIs. Our approach achieved an AUC of 0.74 by training a random forest classifier combining most predictive features. The DL model achieved an AUC of 0.69. This methodology opens the road to interpretability of the response to radiotherapy and tailored treatments for prostate cancer patients.

Keywords: Recurrence prediction · prostate cancer · *in silico* modeling · artificial intelligence · interpretability/explainability

Supplementary Information The online version contains supplementary material available at https://doi.org/10.1007/978-3-031-73376-5_20.

1 Introduction

Prostate cancer is the second most diagnosed cancer in men in the world and the fifth leading cause of death [22]. External Beam Radiotherapy (RT) is the clinical standard treatment for localized prostate cancer [12], which allows to control the tumour in the majority of cases. However, biochemical recurrence (BCR), defined as 2 consecutive elevations of Prostate Specific Antigen (PSA) ≥ 0.2 ng/mL (Phoenix criteria [1]), may occur in 0–10%, 10–20% and 30–40% of patients with respectively low, intermediate and high-risk tumours (according to the D'Amico Classification) within 5 years after the treatment [5]. Predicting BCR prior to RT appears as crucial for assessing patient risk and personalising treatment. Several models have been previously proposed to predict BCR. The first tumour control probability models, [4,14] were based on probability curves, describing the dose-effect relationship on a given population. Nevertheless, these models are limited as they only consider dose discarding the rich nature of the tumour. Recently, radiomics approaches introduced image biomarkers by extracting multiple tumour and organ features from available medical images [3,16], improving performances but facing several issues [6,9]. These approaches require a large amount of training data and they are known to be *black-box* methods lacking interpretability and explainability. Furthermore, they are highly dependent on the data (imbalance classes, image harmonisation, external validation, etc.). Deep learning (DL) models have emerged as an appealing tool to predict BCR, yielding better results than radiomics [17],but sharing the same issues with data. In contrast to these data-driven techniques, mechanistic *in silico* models open new possibilities to predict tumour response by simulating the prescribed irradiation treatment. These models are based on the integration of several biological mechanisms to better understand the response of patients to RT. Their predictive capabilities have been explored in [18]. These models may offer better interpretability and could allow to simulate different patient-specific treatments.

The objective of the present work was thus to propose a novel approach to predict BCR after prostate cancer radiotherapy in a patient specific framework, by combining Magnetic Resonance Imaging (MRI), clinicopathological data, and a radiobiological mechanistic *in silico* model simulating tumour growth and radiation response. The model, based on [20], integrates the most relevant radiobiological mechanisms identified by a sensitivity analysis : oxygenation, tumour cells division and irradiation response. It allows us to create digital twins of patient tumours from MR images and simulate treatments and tissue response. This novel approach was compared to radiomics and deep-learning-based predictions from image and clinicopathological features.

2 Materials and Methods

2.1 Population Dataset

A cohort of 254 patients with localised prostate cancer having undergone RT was used for this study (performed in line with the principles of the Declaration of Helsinki). A detailed description of patient and tumour characteristics can be found in Table 1. Before starting the treatment, 3T MRIs were acquired as described in [9]. It included axial turbo spin echo T2-w and axial diffusion using multiple b-values. N4-bias field correction has been done on T2w images, and apparent diffusion coefficient (ADC) maps were calculated. Prostate and tumour were manually segmented by experts on T2-w sequences and contours were propagated onto the co-registered ADC images (Step (0) in Fig. 1). Patients were followed up through clinical examination and PSA analysis every 6 months for 5 years after the end of irradiation. A total of 39 patients suffered BCR, defined according to the Phoenix criteria [1]. More details about the clinicopathological features available are shown in the Supplementary Materials.

Table 1. Patients description and tumour characteristics.

Patients description		Tumour characteristics (number of instances and their percentage in parenthesis)			
Number of patients		**Pre-treatment PSA (ng/mL)**			
254		$PSA \leq 7$	$7 < PSA \leq 11$	$11 < PSA \leq 20$	$PSA > 20$
Median age (years)		66 (26%)	63 (25%)	66 (26%)	59 (23%)
71		**Clinical stage (T stage)**			
Recurrence		*T1*	*T2*	*T3*	*T4*
Non BCR	**BCR**	23 (9%)	108 (43%)	120 (47%)	3 (1%)
215	39	**Gleason score**			
Total Dose (Gray)		*6*	*7*	*8*	*9*
74–80		38 (15%)	157 (62%)	30 (12%)	29 (11%)

2.2 Mechanisitic *in Silico* model

A previously developed mechanistic *in silico* model simulating tumour growth and response to RT was used [20]. It was implemented in C++, based on the Multiformalism Modeling and Simulation Library (M2SL) [10], allowing the integration of different mechanisms arising at different temporal and spatial scales. It considered 2D digital twins of patient tumours in which each pixel ($20 \mu m$ x $20 \mu m$) represented a cell corresponding to 6 types : healthy (fibroblasts, macrophages, epithelial, muscle, etc.), undamaged tumour, lethally damaged tumour, pre-existing endothelial (vessel cells), neo-created endothelial and dead cells. This mechanistic model integrated major radiobiological mechanisms, occurring at various temporal and spatial dimensions. (a) **Angiogenesis**, the recruitment, creation of new vessels from pre-existing endothelial cells, was based on the diffusion of vascular endothelial growth factor (VEGF). The VEGF distribution was modelled using a reaction-diffusion equation. (b) **Division of healthy and** (c) **tumour cells** based on the cell cycle, which consisted of phases G1 (Gap 1), S (Synthesis), G2 (Gap 2) and M (Mitosis). It also included a fifth phase G0 in which cells were quiescent. (d) **Oxygenation** modeled using a reaction-diffusion equation. (e) **Response to irradiation** given by the linear-quadratic model [8]. To identify the most important parameters and mechanisms involved in tumour response, a sensitivity analysis based Morris screening method was previously performed. Following this sensitivity analysis, a reduced model of 18 parameters was obtained, which included only (c) **Division of tumour cells**, (d) **Oxygenation** and (e) **Response to irradiation**. This model is interpretable in terms of tumour response as it involves understandable equations (differential and linear quadratic equations) and explainable outputs (tumour volume, percentage of cells in the cycle, at each time step). In this paper, we used this reduced mechanistic *in silico* model to simulate the patient specific tumour growth and response to RT. All the equations and parameters details used are fully described in [20].

2.3 Radiomics Feature Extraction

IBSI-compliant [25] features were extracted (Step (1) in Fig. 1) from both modalities within the tumour using the Pyradiomics library [24]. It included 18 first-order statistics, 23 shape descriptors and 66 textural features. Prior to the feature extraction, images were resampled to 1 mm×1 mm×1 mm using BSpline interpolation. This provided a total of 214 features (107 features from T2 modality and 107 from ADC modality). Details about radiomics features extracted are shown in the Supplementary Materials.

2.4 Digital Twin and Patient-Specific *in Silico* simulation

A total of 254 2D digital tissues representing the 254 patients specific tumours of the cohort were built (Step (2) in Fig. 1). These digital twins were initialized using different parameters (i.e. cell size, the tumour area, the tumour density and the vascular density). Every 3D tumour volume was mapped to a disk and considered as a perfect circle (ratio between each axis of the circle equals to 1). The cell size is set to 20.0 reflecting the average size of a cell (i.e. 20.0 μm). The initial area value was computed from the spherical tumour volume. The mean density value was obtained from the average intensity value inside the Volume of Interest (VOI) of the T2-w MR image through linear transformations [15]. These transformations provided 10 supplementary features (5 from T2 and 5 from ADC). An initial prostate-specific vascular density of 3.8% [19] was considered for every virtual tissue (simulating poorly-vascularized tumour core). Then, the prescribed standard irradiation treatment (74–80 Gy administered in 2 Gy fractions from Monday to Friday) was simulated through the 8 weeks of treatment (Step (3) in Fig. 1). The *in silico* simulations produced several outputs at each time point from the beginning of the treatment simulation until 4 weeks after the end of the treatment. Outputs of the mechanistic model were the tumour volume and density, the volume of not damaged tumour cells, the hypoxic cell density, the killed cells percentage, or the percentage of cells in each phase of the cycle.

Prediction of Recurrence After Prostate Cancer Radiotherapy 215

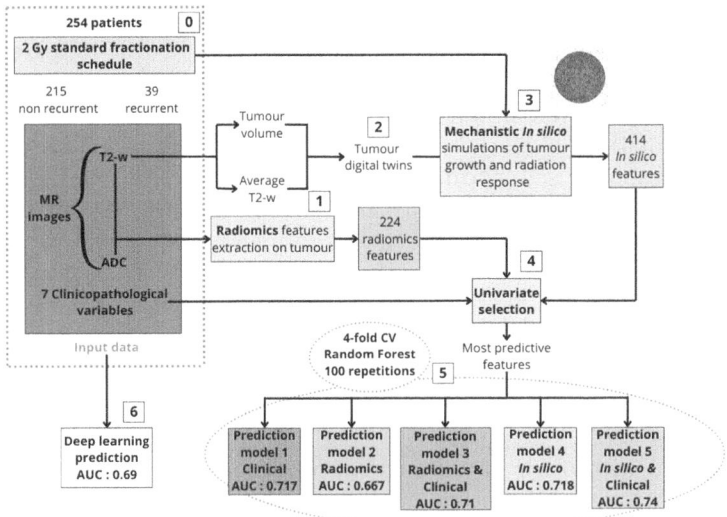

Fig. 1. Workflow. Data preparation and tumour segmentation (0). Image feature extraction (1). Tumour digital twins characterisation and construction (2). *In silico* simulations (3). Univariate feature selection (4). BCR prediction with different feature combinations with AUC prediction scores (5). BCR prediction with deep learning (6).

Exploratory computation of other features, based on *in silico* simulation outputs, was performed in order to produce features that might be good predictors of BCR. Relevant time points were chosen : t = 0; 2; 4; 6; 8 (end of the treatment); 10 and 12 weeks. Values of *in silico* simulation outputs were extracted at each time point, as well as the difference values and the fraction values between each time points. The curve tendency between each time point was also computed. A fitting of each output data was performed between t = 0 until t = 10 weeks (2 weeks after treatment ends) using functions that best fit (for instance, periodic functions for the phases of the cell cycle). Parameters of these fitted curves were extracted and tested as features to predict BCR. Values and times when the fitted function is equal to 5% of the initial output value were also computed. This exploratory feature computation provided 414 features. Details about these mechanistic features are shown in the Supplementary Materials.

2.5 Recurrence Prediction

In total, 645 features (7 clinical, 224 radiomics-based, 414 mechanistic-based) were extracted. A first univariate feature selection analsyis was performed on all these features using a Random Forest Classifier (RFC) stratified 4-fold cross-validation, 20 repetitions (Step (4) in Fig. 1). Through this univariate analysis, the 20 most predictive features were identified (3 clinicopathological, 7 radiomics and 10 mechanistic features). A correlation analysis was performed in order to remove very strongly correlated features (i.e. Spearman's correlation coefficient

> 0.8 [2]). Seven very strongly correlated features were identified (1 clinicopathological, 3 radiomics and 3 mechanistic). This resulted in 13 features split in the different prediction models according to the workflow in Fig. 1. A RFC (stratified 4-fold cross-validation, 100 repetitions) was used to predict BCR with different combinations of features selected (Step (5) in Fig. 1). **Model 1** Only clinical variables were used. **Model 2** Only radiomics features. **Model 3** Model 2 + clinical variables. **Model 4** Only mechanistic features. **Model 5** Model 4 + clinical variables (proposed approach). Performance evaluation was quantified through different metrics : the area under the ROC curve (AUC), the classification accuracy (ACC), the average precision score (APS), the F1 score (F1), the precision (PREC) and the recall (REC).

2.6 Comparison with a Deep Learning Approach

For comparison with other data-based approaches, we developed a DL neural network fed with raw MR images and clinicopathological features to predict binary BCR.

Model Construction. The DL model consisted of 3 parts: (i) A Convolutional Neural Network (CNN) taking as input the cropped T2-w MR images. (ii) A CNN taking as input the cropped ADC maps. The cropped images were centered on the tumour. (iii) A Fully Connected Neural Network (FCNN) taking as input the clinicopathological data. The CNNs (i and ii) used were built with 4 convolutional layers. MaxPooling layers were added to decrease the dimensionality. Flattening layers were inserted at the end of each CNN to produce two vectors of features reflecting the inner images data. The FCNN (iii) used for the clinicopathological data was built with 2 dense layers using L2 regularisation (to prevent overfitting). The outputs of these 3 parts were concatenated as a vector of features and put into another FCNN to predict the binary BCR status. Dense layers were used as in the previous FCNN part. The Sigmoid activation function was applied in the output layer, as the prediction was binary. The rectified linear unit activation function was used in the other layers to overcome the vanishing gradient problem. The dropout regularisation technique was used to prevent overfitting [21]. Batch normalisation layers were added for further regularisation and to reduce the internal covariate shift [11]. The Adam optimiser was used [13]. The number of parameters of this DL model was 25 566 721, reducing model interpretation. **Training and testing** The dataset, consisting in 215 non BCR patients and 39 BCR patients was split with a 75:25% train/test split. During the training, the binary cross-entropy loss was monitored and early stopping was used. As the dataset was imbalanced class weights were applied, defined as :

$$\frac{numberOfPatients}{numberOfClasses * numberOfInstancesOfEachClass}$$

Stratified 4-fold cross-validation, 5 repetitions, was performed. Through the evaluation process, performance was assessed with the same metrics as the other models (Step (6) in Fig. 1).

3 Results and Discussion

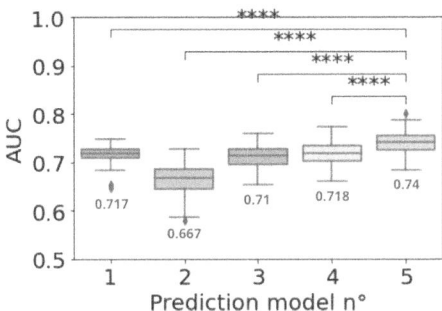

Fig. 2. AUC of prediction models trained with a random forest classifier. Same colour code and model numbers as in Fig. 1 are used. (1) Clinicopathological features; (2) Radiomics features; (3) Radiomics and Clinicopathological features. (4) *In silico* features. (5) *In Silico* and clinicopathological features. The stars correspond to the Wilcoxon signed-rank test significance. **** : p-value $\leq$ 0.0001

The most predictive clinical feature found was the patient ISUP grade group [7]. For the mechanistic-based features, most predictive identified were variables underlying the percentage of tumour cells in phase M such as the curve tendency between 4 and 6 weeks, or the percentage of phase M tumour cells 2 weeks after the end of the treatment. These features are explainable as they have a biological and clinical interpretation. They bring insights into tumour response during RT, that are not visible from any clinical pre-treatment image and, to the best of our knowledge, have never been considered in any predictive model. The most predictive radiomics features identified were the skewness and the texture feature cluster prominence which are more complex to clinically interpret. Univariate feature selection and correlation matrix details are shown in the Supplementary Materials. Results of BCR prediction for each model trained with the RFC are shown in Fig. 2. Results of the model classification performances are shown in Table 2. The proposed approach (Model 5), based on clinical and mechanistic data achieved a mean AUC of 0.74 over the 100 repetitions, outperforming the four other models trained with the other feature combinations. It was significantly better than these models (p-value $\leq$ 0.0001, given by a Wilcoxon signed-rank test with Bonferroni correction). It was not significantly better than models 2 and 3 in terms of precision. The DL model achieved a mean AUC of 0.69, not outperforming the proposed approach in terms of AUC, accuracy and average precision. However, in terms of F1 score and recall, the proposed novel approach was outperformed by the DL model and the radiomics ones.

This study presents several limitations to be explored in future work. First of all, representing tumours via spherical mapping may seem simplistic, however this was also used by [19], and simulation results were in good agreement with their *in*

Table 2. BCR prediction performance. Bold denotes top or significantly close to top results (Wilcoxon signed-rank test shows no significant difference)

Model	ACC	APS	F1	PREC	REC	AUC
Random Forest Classifier						
Model 1 (Clinical Data)	0.837	0.320	0.033	0.029	0.018	0.717
Model 2 (Radiomics Data)	0.840	0.376	0.237	**0.491**	0.174	0.667
Model 3 (Radiomics + Clinical)	0.839	0.389	0.233	**0.451**	0.167	0.710
Model 4 (Mechanistic Data)	0.830	0.325	0.064	0.194	0.004	0.718
Model 5 (Mechanistic + Clinical)	**0.848**	**0.418**	0.198	**0.494**	0.133	**0.740**
Deep Learning model	0.834	0.338	**0.476**	**0.497**	**0.475**	0.690

AUC: area under the ROC curve; ACC: classification accuracy;
APS: average precision score; F1: F1 score; PREC: precision; REC: recall

vivo observations in mice. Initialisation of digital twins of patient tumours could have included ADC maps, which may provide additional information on cell density. The use of IntraVoxel Incoherent Motion DW-MRI could also be explored to better initialise patient-specific digital twins. Secondly, the mechanistic *in silico* model used has to be enriched with other biological mechanisms, such as the immune response. The model can be validated with animal experiments, as so far all the parameter values are set from the literature. Other parameters, such as the vascular density (which was set to a constant value), have to be calibrated properly from other image modalities. This could be done by assessing the hypoxia in the tumour which is well-known as a critical prognosis factor related to radio resistance and recurrence [23]. Furthermore, the exploratory feature selection can be improved using other strategies, such as dimensional reduction. The results presented in Table 2 may seem underwhelming, in terms of F1 and recall, this is mainly due to the highly imbalanced nature of the dataset. Artificial oversampling techniques, such as SMOTE, can be included in this work. Finally, the approach has to be validated on an external population.

4 Conclusion and Future Work

We proposed a novel approach to predict BCR after prostate cancer RT. It was based on the combination of clinical features and an *in silico* model of tumour growth and radiation response, initialised with MRI-based digital twins of tumours. This approach brings interpretability and explainability to the prediction, outperforming models based on clinical data, radiomics or DL techniques. Hence, this method enables to personalise treatments by stratifying patients with low or high risk of recurrence, but also opens the possibility to simulate tailored plannings.

Acknowledgments. With financial support from ITMO Cancer of Aviesan within the framework of the 2021–2030 Cancer Control Strategy, on funds administered by Inserm.

Disclosure of interests. The authors have no competing interests to declare that are relevant to the content of this article.

References

1. Abramowitz, M.C., et al.: The phoenix definition of biochemical failure predicts for overall survival in patients with prostate cancer. Cancer **112**(1), 55–60 (2008)
2. Akoglu, H.: User's guide to correlation coefficients. Turkish J. Emerg. Med. **18**(3), 91–93 (2018)
3. Avanzo, M., et al.: Machine and deep learning methods for radiomics. Med. Phys. **47**(5), e185–e202 (2020)
4. Chanrion, M.-A., et al.: The influence of the local effect model parameters on the prediction of the tumor control probability for prostate cancer. Phys. Med. Biol. **59**(12), 3019 (2014)
5. D'Amico, A.V., et al.: Biochemical outcome after radical prostatectomy, external beam radiation therapy, or interstitial radiation therapy for clinically localized prostate cancer. JAMA **280**(11) (1998)
6. Duenweg, S.R., et al.: T2-weighted MRI radiomic features predict prostate cancer presence and eventual biochemical recurrence. Cancers **15**(18), 4437 (2023)
7. Epstein, J.I., et al.: The 2014 International Society of Urological Pathology (ISUP) consensus conference on gleason grading of prostatic carcinoma: definition of grading patterns and proposal for a new grading system. Am. J. Surg. Pathol. **40**(2), 244 (2016)
8. Fowler, J.F.: The linear-quadratic formula and progress in fractionated radiotherapy. British J. Radiol. **62**(740), 679–694 (1989)
9. Gnep, K., et al.: Haralick textural features on T_2-weighted MRI are associated with biochemical recurrence following radiotherapy for peripheral zone prostate cancer: impact of MRI in prostate cancer. J. Magn. Reson. Imaging **45**(1), 103–117 (2017)
10. Hernández, A.I., et al.: A multiformalism and multiresolution modelling environment: application to the cardiovascular system and its regulation. Philos. Trans. Royal Soc. A: Math. Phys. Eng. Sci. **367**(1908), 4923–4940 (2009)
11. Ioffe, S., Szegedy, C.: Batch normalization: accelerating deep network training by reducing internal covariate shift. In: International Conference on Machine Learning, pp. 448–456. PMLR (2015)
12. Joiner, M.C., van der Kogel, A.J. (eds.) Basic Clinical Radiobiology. CRC Press (2018)
13. Kingma, D.P., Ba, J.: Adam: a method for stochastic optimization (2014)
14. Kupelian, P.A., et al.: Effect of increasing radiation doses on local and distant failures in patients with localized prostate cancer. Int. J. Radiat. Oncol. Biol. Phys. **71**(1), 16–22 (2008)
15. Kwak, J.T., et al.: Prostate cancer: a correlative study of multiparametric MR imaging and digital histopathology. Radiology **285**(1), 147–156 (2017)
16. Lambin, P., et al.: Radiomics: extracting more information from medical images using advanced feature analysis. Eur. J. Cancer **48**(4), 441–446 (2012)
17. Lee, H.W., et al.: Novel multiparametric magnetic resonance imaging- based deep learning and clinical parameter integration for the prediction of long-term biochemical recurrence-free survival in prostate cancer after radical prostatectomy. Cancers **15**(13), 3416 (2023)

18. Nicolò, C., et al.: Machine learning and mechanistic modeling for prediction of metastatic relapse in early-stage breast cancer. JCO Clin. Cancer Inform. **4**, 259–274 (2020)
19. Paul-Gilloteaux, P., et al.: Optimizing radiotherapy protocols using computer automata to model tumour cell death as a function of oxygen diffusion processes. Sci. Rep. **7**(1), 2280 (2017)
20. Sosa-Marrero, C., et al.: Towards a reduced in silico model predicting biochemical recurrence after radiotherapy in prostate cancer. IEEE Trans. Biomed. Eng. **68**(9), 2718–2729 (2021)
21. Srivastava, N., et al.: Dropout: a simple way to prevent neural networks from overfitting. J. Mach. Learn. Res. **15**(1), 1929–1958 (2014)
22. Sung, H., et al.: Global cancer statistics 2020: GLOBOCAN estimates of incidence and mortality worldwide for 36 cancers in 185 countries. CA: Cancer J. Clin. **71**(3), 209–249 (2021)
23. Tatum, J.L.: Hypoxia: importance in tumor biology, noninvasive measurement by imaging, and value of its measurement in the management of cancer therapy. Int. J. Radiat. Biol. **82**, 699–757 (10) (2006)
24. Van Griethuysen, J.J., et al.: Computational radiomics system to decode the radiographic phenotype. Cancer Res. **77**(21), e104–e107 (2017)
25. Zwanenburg, A., et al.: The image biomarker standardization initiative: standardized quantitative radiomics for high-throughput image-based phenotyping. Radiology **295**(2), 328–338 (2020)

AutoDoseRank: Automated Dosimetry-Informed Segmentation Ranking for Radiotherapy

Zahira Mercado[1], Amith Kamath[1], Robert Poel[1,3], Jonas Willmann[2], Ekin Ermis[3], Elena Riggenbach[3], Lucas Mose[3], Nicolaus Andratschke[2], and Mauricio Reyes[1,3(✉)]

[1] ARTORG Center for Biomedical Engineering Research, University of Bern, Bern, Switzerland
[2] Department of Radiation Oncology, University Hospital Zurich, University of Zurich, Zurich, Switzerland
[3] Department of Radiation Oncology, Inselspital, Bern University Hospital and University of Bern, Bern, Switzerland
mauricio.reyes3@unibe.ch

Abstract. AutoDoseRank (Automated Dosimetry-informed Segmentation Ranking) is a novel methodology for ranking segmentations of glioblastoma, a highly aggressive brain tumor, by dosimetric quality. AutoDoseRank uses a deep learning-based dose predictor along with a dosimetric ranking scheme capable of sorting a set of candidate segmentations by quality. With the advent of auto-segmentation for radiotherapy, we expect that radiation oncologists will spend more time triaging and evaluating the quality of generated segmentation proposals rather than manually drawing them. It is known that changes in segmentation evaluated purely through a geometric lens like Dice, do not correlate with eventual clinical outcomes. Our approach therefore aims to incorporate organ-specific dosimetric constraints used in clinical radiotherapy planning into a patient-level ranking. The effectiveness of AutoDoseRank is measured by comparing its ability to rank segmentations against that of four expert radiation oncologists. We show that AutoDoseRank is better than three out of four experts while being only slightly outperformed by the most experienced and meticulous one. These results highlight AutoDoseRank's capability to monitor the quality of auto-segmentation dosimetrically, something that is ever increasing in importance in the radiation oncology workflow. Code to reproduce this analysis is available at https://github.com/ubern-mia/autodoserank.

Keywords: Segmentation quality · Radiotherapy · Tumor control · Tissue toxicity

1 Introduction

Glioblastoma is a particularly aggressive brain tumor that constitutes approximately 45% of all brain tumors [11]. The conventional treatment strategy includes surgery, radiotherapy (RT), and chemotherapy, with the objective of precisely targeting the tumor while sparing as much healthy tissue (also known as organs-at-risk (OAR)) as feasible [18]. This necessitates a careful equilibrium between tumor coverage and tissue toxicity control. An integral component of this procedure is the segmentation of structures, which can be laborious when performed manually [4]. The entire process, from the patient's initial scan to the commencement of the RT treatment, can take more than a week on average, with segmentation and RT planning being the most time-intensive stages [5]. The advent of deep learning-based auto-segmentation is anticipated to significantly decrease this time.

This suggests that the role of medical professionals will transition from performing the segmentation task to supervising and rectifying the outcomes generated by automated systems. The task of ensuring the quality of an automated segmentation is crucial, as errors in tumor segmentation have been shown to cause 25% of treatment plans to be non-compliant, potentially leading to untreated tumors or harmful radiation doses [12]. It's crucial to emphasize that in radiation therapy, the impact of segmentation variations on the administered dose is a determining factor in clinical outcomes as it influences tumor control and toxicity to healthy tissue. There is an urgent need for dosimetric considerations in the process of assessing and measuring the effects of corrected or alternative automated segmentations [2].

In the context of computer vision and natural image scenarios, geometric measures such as the Dice score coefficient (DSC) and Hausdorff distance are frequently employed to assess the quality of segmentations. These measures have also been utilized to automate the evaluation of segmentation quality [3,6,19]. However, these metrics do not strongly correlate with either the dosimetric consequences of contouring errors [14] or the assessment of clinicians [10]. Recent studies [15,21] further underscore the importance of incorporating domain knowledge when selecting and designing metrics.

The integration of dosimetric effects into the evaluation process calls for the computation of dose plans. These computations are time-consuming and necessitate a collaborative effort between the radiation oncologist and dosimetrists or medical physicists. Due to its time-intensive nature, dose-guided evaluation of segmentation quality has not been commonly employed in clinical environments.

With recent advances in deep learning-based models for radiotherapy planning, a dose prediction model has been used to guide radiation oncologists on the volume slices that require manual modifications [16]. This segmentation editing tool demonstrates potential for time efficiency while maintaining dosimetric equivalence with dose distribution maps produced without its assistance. Furthermore, methods that utilizes a deep learning-based dose predictor to assess the impact of local segmentation alterations on dosimetric outcomes have also been described [8]. However, it mainly concentrates on OAR segmentation and not on tumor lesions, which hold more clinical significance due to the increased variability in the segmentation task. Moreover, these methods do not offer a technique for ranking different segmentation candidates or alternatives based on their dosimetric effects, which is of higher clinical interest to assist medical experts in the supervision and correction of automated segmentation results. These could potentially lead to more precise and efficient treatment plans.

Contributions: We propose the first (to the best of our knowledge) dose-informed framework for ranking a set of segmentations, taking advantage of recent advancements in deep learning-based dose prediction. This dosimetric triage of segmentations is based on (i) OAR-specific dose constraints and (ii) relative prioritization between OARs. This effectively brings forward into the workflow of evaluating segmentation quality, knowledge from the subsequent step of RT planning for brain tumor patients, thereby making the process more clinically relevant. We demonstrate the ability of the proposed approach, termed AutoDoseRank for **Auto**mated **Do**simetry-informed **Seg**mentation **Rank**ing, by comparing and analyzing its performance against four radiation oncologists.

2 Methods

AutoDoseRank is made up of two primary elements: (i) a deep learning-based dose predictor to estimate dose distributions for a segmentation candidate; and (ii) a ranking module that examines a set of such segmentations based on clinically interpretable parameters, including OAR specific dose effects and planning prioritization. These components together help construct clinically relevant segmentation candidate rankings.

Formally, given an initial original segmentation S_{orig} and corresponding segmentation candidates $\{S_a\}_{a:1,...,n}$, corresponding dose predictions, D_{orig} and $\{D_a\}_{a:1,...,n}$ are computed using a deep learning-based dose predictor model $DP(S) \to D \in \mathbb{R}^{W \times H \times C}$, for the original and candidate segmentations, respectively. In practice, the candidate segmentations could be generated as part of an expert's online correction workflow or come as proposals from an auto-segmentation model. Each voxel in D corresponds to the local predicted dose in Gray (Gy). We utilize a previously reported model [7], with an average dose scores of 1.38 Gy [13]. The novelty in AutoDoseRank beyond using dose predictors to inform quality assessments to help rank and triage candidate segmentations is a new decomposable patient-level metric for estimating dose impact, DI. DI can be decomposed into per-OAR impact, DI_i, computed using the predicted dose map $\tilde{D}_a = D_a \cdot m_i$ and the original dose map $\tilde{D}_{orig} = D_{orig} \cdot m_i$, where m_i is the mask for the i-th OAR.

$$DI_i = \frac{\Phi(\tilde{D}_a) - \Phi(\tilde{D}_{orig})}{C_i} \left(\frac{\Phi(\tilde{D}_{orig})}{C_i}\right)^\gamma \qquad \textit{OAR level} \qquad (1)$$

$$DI = DI(S_a) = \frac{\sum_i \frac{1}{p_i} DI_i}{\sum_i \frac{1}{p_i}} \qquad \textit{patient level} \qquad (2)$$

D_{orig} corresponds to the dose map of the original segmentation S_{orig}. $\Phi(D)$ represents either the maximum or mean dose computed from the dose map D

within the boundaries of the OAR [13]. C_i corresponds to the dose constraint for the i-th OAR, complying with current clinical guidelines. The parameter γ enables regulating the penalization at the OAR level when the candidate alternative dose metric approaches the OAR-specific dose constraint. In our experiments, parameter $\gamma = 1$ for a linear penalization at the patient level. Finally, p_i corresponds to the priority of the i-th OAR, which is specified based on clinical recommendations and used within the treatment planning system (TPS) (e.g., p("eye")=8, p("optic nerve")=1, etc.).

Dose impact values, denoted as $DI(S_a)$, are obtained for each candidate segmentation. These values are then ranked in ascending order to establish the final ranking. The intuition behind equation (1) and (2) is to penalize dose variations caused by a candidate segmentation. This is relative to the OAR-specific constraint (i.e., Eq. (1) OAR level) further intensified when the candidate dose metric approaches the corresponding OAR specific dose constraint.

2.1 Data

Our evaluation data set comprises 65 segmentation candidates from 13 Glioblastoma patients. This dataset includes CT scans, T1 contrast-enhanced MR images, and binary segmentation masks of 13 OARs, as well as the tumor target volume. The OARs consist of the brainstem, optic chiasm, cochleae, eyes, hippocampi, lacrimal glands, optic nerves, and the pituitary gland. In addition, each case also includes a ground truth dose plan, computed with Eclipse (Varian Medical Systems Inc., Palo Alto, USA), and a predicted dose plan using a cascaded 3D UNet (trained separately on 60 independent patients) [7]. The ground truth is computed using a standard clinical protocol with double arc co-planar volumetric modulated arc therapy (VMAT) plan with 6 mega volt flattening filter-free beams. It is optimized (using Varian photon optimizer version 15.6.05) to deliver 30 fractions of 2 Gy each, while sparing the OARs as much as possible. The dose is calculated using the AAA algorithm [20], and normalized such that the prescribed dose fully covers 50% of the target volume. All volumes are resampled to an isometric $2 \times 2 \times 2$ millimeter grid of size 128^3 voxels using PyRaDiSe [17] and converted to NIfTI files for training and evaluation purposes.

2.2 Experimental Setup

For each of the 13 patients, an experienced RT planner manually generated four distinct segmentation candidates, leading to a total of 65 segmentations (i.e., 13 original segmentations $+13 \times 4$ candidates $= 65$ segmentations), on which dose plans are calculated with Eclipse (Varian Medical Systems Inc., Palo Alto, USA). Each original segmentation is ranked per case using the dose impact formulation (Eq. 1) using Eclipse (clinical TPS) dose map distributions instead of the deep learning-based dose predictions. Both the candidate modifications and the creation of RT plans are time-consuming and effort-intensive tasks. Figure 1 shows an example of a original and corresponding segmentation candidates.

Expert-Based Ranking of Segmentation Candidates: We perform evaluation sessions for each test case with four experienced radiation oncologists, who are asked to rank the four candidates based on their potential negative impact on the dose received by the OARs. We use 3D Slicer (v5.6.0) to display all three slice planes. We also provide a 3D view of the spatial relationship between the OARs and the tumor target volume, highlighting the location of the candidate alteration. All variations for each subject are presented simultaneously, allowing for visual comparison against the original.

Evaluation: We adopt the following metrics, commonly used to evaluate rankings: (1) The Normalized Distance-based Performance Measure (NDPM), a ranking metric that quantifies the accuracy of a ranking [1]. NDPM yields a value in the range of $[0, 1]$, where 0 indicates an ideal ranking and 1 indicates the maximum deviation from the ideal ranking. Essentially, NDPM imposes a penalty on a ranking proportional to its deviation from the perfect ranking (2). Kendall's Tau is a statistical measure used to quantify the ordinal association between two measured quantities [9]. It is a non-parametric test that measures rank correlation, or the similarity of the orderings of the data when ranked by each of the quantities. The value of Kendall's Tau is in the range of $[-1, 1]$, where a value of 1 indicates the ranking's complete agreement, -1 indicates the rankings are the exact reverse of each other, and a value of 0 indicates no relationship. Kendall's Tau can provide valuable insights into the relationship between two sets of rankings. We also perform a bootstrapping analysis of the Kendall Tau ranking correlations to obtain further insights into the rankings under sampling variability. We report results performing 1000 sampling with replacement and 90% confidence intervals (CI).

3 Results

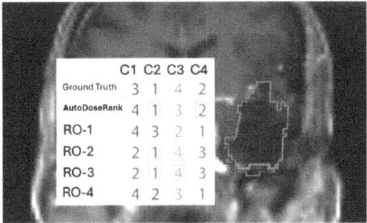

Fig. 1. A representative example of original tumor target volume segmentation (green outline) with four candidates denoted as: C1 (red), C2 (dark blue), C3 (turquoise) and C4 (purple). The table inlaid shows how Ground Truth (Eclipse), AutoDoseRank, and four experts (RO-1 to 4) rank these in order of dose impact. Yellow boxes indicate correct matches with the ground truth. (Color figure online)

Figure 1 presents a typical instance of a case that includes the original segmentation, four alternative candidates, and the corresponding rankings given by AutoDoseRank and four experts. In general, the experts vary significantly in time taken to complete the ranking task, from 19 to 138 s per candidate. AutoDoseRank on the other hand takes a constant time quantum (of ~ 30 seconds), regardless of geometrical complexity.

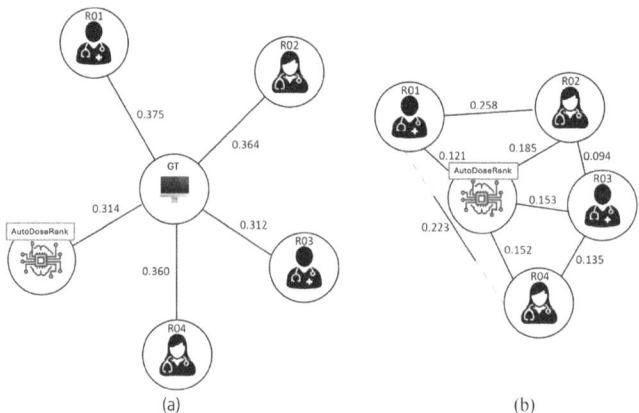

Fig. 2. Normalized Distance-based Performance Measure (NDPM) visualization with scaled distance lengths representing value: (a) NDPM compared to Eclipse (ground truth) with a scaled distance of 0.75 (b) Cross-references between AutoDoseRank and the radiation oncologists with scaled distance of 1. Note: RO1 vs RO4 shown with a dotted line; black continuous line indicates correct scale.

Figure 2 shows the Normalized Distance-based Performance Measure (NDPM) as a graph. This metric compares Eclipse, AutoDoseRank, and each of the four experts. A lower NDPM value indicates a higher level of agreement. As shown in Fig. 2(a), AutoDoseRank and expert "RO-3" exhibit similar ranking performance (two lowest NDPM), outperforming the other three experts. Figure 2(b) presents NDPM within the group of experts and AutoDoseRank, showing a higher level of agreement among them compared to their agreement with the ground truth (Eclipse). Furthermore, AutoDoseRank's agreement with the experts is more consistent than the inter-rater agreement, indicating that it provides more reliable rankings.

Table 1 shows Kendall's Tau ranking correlation summary performed with 1000 resamplings comparing AutoDoseRank and the four radiation oncologists to the ground truth (Eclipse). AutoDoseRank shows a stronger correlation (between weak to moderate) to the ground truth rankings and outperforms three out of four experts. Interestingly, AutoDoseRank performs similarly to RO-3, who is the most experienced and meticulous amongst the four.

Table 1. Summary of Kendall's Tau ranking correlation performed with 1000 resamplings comparing AutoDoseRank and the four experts, denoted as RO-1 to -4, to the ground truth. RO-3 and AutoDoseRank yield higher correlations ranging from weak to moderate, outperforming the others. RO-3 is the most experienced and meticulous expert, who also took the longest time to perform the task.

Ground Truth (Eclipse) vs	Mean Kendall's Tau	90% CI
AutoDoseRank	0.129	[-0.097, 0.354]
RO-1	0.014	[-0.194, 0.231]
RO-2	0.038	[-0.163, 0.239]
RO-3	0.148	[-0.056, 0.347]
RO-4	0.041	[-0.161, 0.238]

To complement Table 1, and obtain more insights into the distribution of Kendall's Tau ranking correlations, Fig. 3 presents the Cumulative Distribution Functions (CDF) of Kendall's Tau metric values, calculated over 1000 bootstrapping resamplings for AutoDoseRank and the four experts. The distribution of these values corroborates the NDPM findings, indicating that AutoDoseRank and expert 'RO-3' outperform the other three experts. This superior performance is highlighted by the noticeable gap between their CDF curves in Fig. 3.

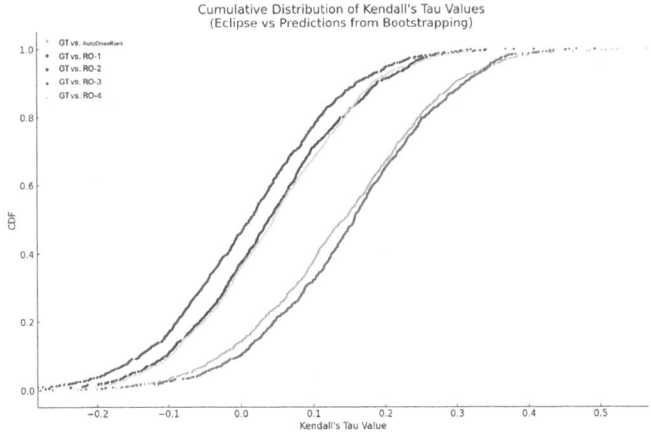

Fig. 3. Comparative Cumulative Distribution Functions of Kendall's Tau Correlation Coefficients: Ground Truth versus AutoDoseRank and four experts.

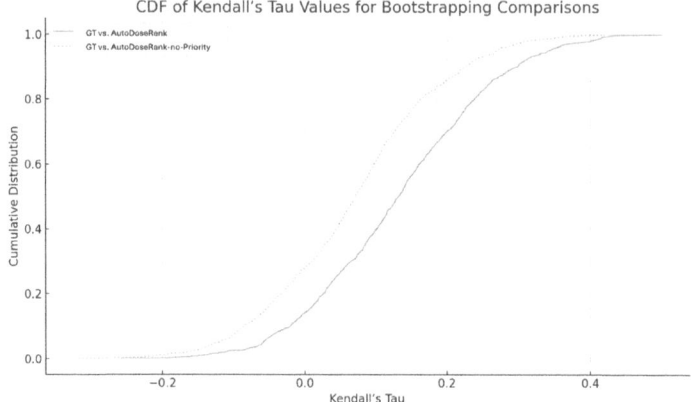

Fig. 4. Ablation on OAR prioritization: Comparative Cumulative Distribution Functions of Kendall's Tau Correlation Coefficients: The ground truth versus AutoDoseRank with and without priority weighting.

Ablation Experiment on Priority: Fig. 4 shows the Cumulative Distribution Function of Kendall's Tau Correlation coefficients of AutoDoseRank with and without applying different priorities to each OAR. This assesses the importance of prioritization used when computing the patient-level dose impact. We observe a clear worsening in the ranking with no prioritization, denoted as AutoDoseRank-No-Priority.

4 Discussion and Conclusion

In this paper, we present AutoDoseRank, a dose-informed deep learning-based approach to rank segmentation candidates, aiming at aiding radiation oncologists in the task of monitoring segmentation quality. AutoDoseRank outperformed three out of four experts while being slightly outperformed by the most experienced and meticulous one. This highlights the capability of this framework to assist clinicians in monitoring segmentation quality semi-autonomously.

Bearing in mind that the clinician's time for running these human versus automated-model evaluations is not easily scalable, we emphasize the multi-expert evaluation presented, along with some limitations. Using more than 13 clinical cases to evaluate the approach could certainly improve the statistical robustness of our findings. The segmentation and planning preparation process consumes more than 70% of the total time from diagnosis to treatment, which typically spans over two weeks [5]. Consequently, producing meaningful segmentation candidates and re-planning corresponding dose maps using Eclipse (clinical TPS) is a significant time investment. We aim to mitigate this by building more accurate dose prediction models. Such models could also help us evaluate the sensitivity (by modifying γ) of the analysis to dose changes. Furthermore, we plan to run these evaluations with more radiation oncologists and dosimetrists to assess how experts vary amongst themselves versus future iterations of AutoDoseRank. Beyond the presented results, we believe this framework can be used in training radiation oncologists to improve their dosimetric awareness to segmentation variations.

References

1. Avazpour, I., Pitakrat, T., Grunske, L., Grundy, J.: Dimensions and metrics for evaluating recommendation systems. In: Robillard, M.P., Maalej, W., Walker, R.J., Zimmermann, T. (eds.) Recommendation Systems in Software Engineering, pp. 245–273. Springer Berlin Heidelberg, Berlin, Heidelberg (2014). https://doi.org/10.1007/978-3-642-45135-5_10
2. Claessens, M., et al.: Quality assurance for AI-Based applications in radiation therapy. In: Seminars in Radiation Oncology, vol. 32, pp. 421–431 (2022). https://doi.org/10.1016/j.semradonc.2022.06.011
3. Cubero, L., Serrano, J., Castelli, J., De Crevoisier, R., Acosta, O., Pascau, J.: Exploring uncertainty for clinical acceptability in head and neck deep learning-based OAR segmentation. In: Proceedings - International Symposium on Biomedical Imaging, vol. 2023-April. IEEE Computer Society (2023).https://doi.org/10.1109/ISBI53787.2023.10230442
4. Das, I.J., Moskvin, V., Johnstone, P.A.: Analysis of treatment planning time among systems and planners for intensity-modulated radiation therapy. J. Am. Coll. Radiol. **6**(7), 514–517 (2009). https://doi.org/10.1016/j.jacr.2008.12.013
5. Guo, C., Huang, P., Li, Y., Dai, J.: Accurate method for evaluating the duration of the entire radiotherapy process. J. Appl. Clin. Med. Phys. **21**(9), 252–258 (2020)
6. Henderson, E.G.A., Green, A.F., Van Herk, M., Osorio, E.M.V.: Automatic identification of segmentation errors for radiotherapy using geometric learning (2022). http://arxiv.org/abs/2206.13317
7. Kamath, A., Poel, R., Willmann, J., Andratschke, N., Reyes, M.: How sensitive are deep learning based dose prediction models to variability in organs at risk segmentation? In: International Symposium on Biomedical Imaging. pp. 1–4 (2023)
8. Kamath, A., Poel, R., Willmann, J., Ermis, E., Andratschke, N., Reyes, M.: ASTRA: atomic surface transformations for radiotherapy quality assurance. In: 45th Annual International Conference of the IEEE Engineering in Medicine & Biology Society (EMBC). Sydney (2023).https://doi.org/10.1109/EMBC40787.2023.10341062. https://github.com/amithjkamath/astra
9. Kendall, M.G.: A new measure of rank correlation. Biometrika **30**(1/2), 81–93 (1938)
10. Kofler, F., et al.: Are we using appropriate segmentation metrics? Identifying correlates of human expert perception for CNN training beyond rolling the DICE coefficient. J. Mach. Learn. Biomed. Imag. **2023**, 27–71 (2023). https://melba-journal.org/2023:002
11. McFaline-Figueroa, J.R., Lee, E.Q.: Brain tumors (2018).https://doi.org/10.1016/j.amjmed.2017.12.039
12. Peters, L.J., et al.: Critical impact of radiotherapy protocol compliance and quality in the treatment of advanced head and neck cancer: results from TROG 02.02. J. Clin. Oncol. **28**(18), 2996–3001 (2010). https://doi.org/10.1200/JCO.2009.27.4498
13. Poel, R., et al: Deep-learning-based dose predictor for glioblastoma-assessing the sensitivity and robustness for dose awareness in contouring. Cancers **15**(17) (2023).https://doi.org/10.3390/cancers15174226

14. Poel, R., et al.: The predictive value of segmentation metrics on dosimetry in organs at risk of the brain. Med. Image Anal. **73**, 102161 (2021). https://doi.org/10.1016/j.media.2021.102161.
15. Reinke, A., et al.: Understanding metric-related pitfalls in image analysis validation. Nat. Methods **21**(2), 1–13 (2024)
16. Roberfroid, B., Lee, J.A., Geets, X., Sterpin, E., Barragán-Montero, A.M.: DIVE-ART: a tool to guide clinicians towards dosimetrically informed volume editions of automatically segmented volumes in adaptive radiation therapy. Radiother. Oncol. **192**, 110108 (2024).https://doi.org/10.1016/j.radonc.2024.110108. https://linkinghub.elsevier.com/retrieve/pii/S016781402400029X
17. Rüfenacht, E., et al.: PyRaDiSe: a python package for DICOM-RT-based auto-segmentation pipeline construction and DICOM-RT data conversion. Comput. Methods Programs Biomed. **231** (2023). https://doi.org/10.1016/j.cmpb.2023.107374
18. Stupp, R., et al.: Radiotherapy plus concomitant and adjuvant temozolomide for glioblastoma. New Engl. J. Med. **352**(10), 987–996 (2005). www.nejm.org
19. Valindria, V.V., et al.: reverse classification accuracy: predicting segmentation performance in the absence of ground truth (2017). http://arxiv.org/abs/1702.03407
20. Van Esch, A., et al.: Testing of the analytical anisotropic algorithm for photon dose calculation. Med. Phys. **33**(11), 4130–4148 (2006). https://doi.org/10.1118/1.2358333
21. Wang, Z., Wang, E., Zhu, Y.: Image segmentation evaluation: a survey of methods. Artif. Intell. Rev. **53**, 5637–5674 (2020)

SurvCORN: Survival Analysis with Conditional Ordinal Ranking Neural Network

Muhammad Ridzuan(✉)[iD], Numan Saeed[iD], Fadillah Adamsyah Maani[iD], Karthik Nandakumar[iD], and Mohammad Yaqub[iD]

Mohamed Bin Zayed University of Artificial Intelligence, Abu Dhabi, UAE
{muhammad.ridzuan,numan.saeed,fadillah.maani,
karthik.nandakumar,mohammad.yaqub}@mbzuai.ac.ae

Abstract. Survival analysis plays a crucial role in estimating the likelihood of future events for patients by modeling time-to-event data, particularly in healthcare settings where predictions about outcomes such as death and disease recurrence are essential. However, this analysis poses challenges due to the presence of censored data, where time-to-event information is missing for certain data points. Yet, censored data can offer valuable insights, provided we appropriately incorporate the censoring time during modeling. In this paper, we propose SurvCORN, a novel method utilizing conditional ordinal ranking networks to predict survival curves directly. Additionally, we introduce SurvMAE, a metric designed to evaluate the accuracy of model predictions in estimating time-to-event outcomes. Through empirical evaluation on two real-world cancer datasets, we demonstrate SurvCORN's ability to maintain accurate ordering between patient outcomes while improving individual time-to-event predictions. Our contributions extend recent advancements in ordinal regression to survival analysis, offering valuable insights into accurate prognosis in healthcare settings. Our code is available at https://github.com/BioMedIA-MBZUAI/SurvCORN.

Keywords: survival analysis · prognosis · ordinal ranking

1 Introduction

Survival analysis estimates the probability of a future event occurring to patients by modeling time-to-event data. In healthcare settings, common medical applications include predicting the time to death, recurrence of diseases, or re-hospitalization of patients using medical images and Electronic Health Records (EHRs). Survival analysis is a challenging problem due to the presence of censored data; for certain data points, the time-to-event information is missing due to various reasons such as patients discontinuing follow-up visits, relocating, or withdrawing from a study [22]. However, censored data can be useful if we utilize the censoring time during modeling because it entails a time until which we are sure that the event did not happen, e.g., the patient did not die.

A commonly reported metric in survival analysis is the concordant index, which is used to evaluate the pairwise concordance of survival times [8–10]. However, it does not provide a simple, interpretable assessment of the actual time-to-event predictions. Patients and clinicians alike are likely to benefit from saying a model has a prediction error of X number of days than simply saying it has a concordant index of Y.

Deep survival methods can be broadly categorized into continuous-time and discrete-time methods, where the discrete methods approximate the continuous models by discretizing the continuous survival time scale. Continuous methods include DeepSurv [14], built upon the semi-parametric Cox Proportional Hazard (CoxPH) [5,14] model which assumes the ratio of the hazard functions for any two individuals is constant over time. Discrete methods include Nnet-Survival [7] and DeepHit [17]. DeepHit [17] parameterizes the event-time probability mass function, while Nnet-Survival parameterizes the hazard rate using a Bernoulli function. Our proposed method falls under the discrete category.

A pivotal tool in survival analysis is the survival curve, offering a graphical representation of the fraction of patients surviving after a specific event. The survival curve is a decreasing function that allows healthcare professionals to assess the probability of survival over time and compare survival experiences between different patient groups or treatment modalities. Recently, Shi et al. [21] introduced a rank-consistent ordinal regression for neural networks based on conditional probabilities. They provide strong theoretical guarantees for rank-monotonicity, where the rank of an object changes monotonically with its predicted probability of belonging to a higher- or lower-ranked category.

We observe this property to be desired for survival analysis, where the predicted probability of a patient's survival decreases over time in the absence of an intervention. We thus propose an extension to this method that accounts for both censored and uncensored cases and directly predicts the survival curve using a conditional probability interpretation of the network output. Our contributions are two-fold:

– We introduce *SurvCORN*, a *Surv*ival analysis method using *C*onditional *O*rdinal *R*anking *N*etwork, that directly predicts patients' survival curves using conditional probabilities
– We propose *SurvMAE*, a metric that evaluates the quality of a model's survival predictions based on how far they are from the actual recorded time-to-events

We show empirically that SurvCORN is able to maintain a correct ordering of patient outcomes while improving upon individual time-to-event predictions.

2 Method

We aim to develop a deep neural network that directly predicts a monotonically decreasing function from $1 \rightarrow 0$ to represent a patient's survival probability over time. We first divide the time axis into K number of discrete intervals, or *time bins*, then use logistic regression to predict the patient's survival probability beyond each time bin. The K-th time bin represents a final all-encompassing time frame beyond the maximum time in the training set (T_{max}, ∞), i.e., right-censored at T_{max}.

Given a training dataset $D = \{X^i, \delta^i, T^i\}_{i=1}^N$, where N is the number of patients, X^i is the patient features (e.g. MRI, CT scans, EHR), δ^i is the event indicator (with $\delta^i = 1$ indicating event occurrence for uncensored patients, while $\delta^i = 0$ indicating event unobserved for censored patients), and T^i is the actual time-bin index of the discretized time-to-event. The objective is to train a network $f_\theta(X) \to z \in \mathbb{R}^{d_{K-1}}$ to predict a patient's survival probability beyond each time bin t_k, denoted as $S(T > t|X) = P(T > t \mid X)$. The output of the network is of size $K - 1$, not K, because a patient who survives beyond t_{K-1} is assumed to experience the event at t_K. In other words, a high probability of surviving beyond t_{K-1} means that the patient will likely experience the event at t_K.

2.1 Label Encodings

Unique to survival analysis is the presence of censoring. Here, we present a separate label encoding for uncensored versus censored patients that is crucial for the minimization of the log-likelihood.

Uncensored Encoding ($\delta^i = 1$). For uncensored patients, the time bin index T^i is transformed into an ordered vector representation of $K - 1$ binary labels $\{T_1^i, T_2^i, \ldots, T_{K-1}^i\}$, where $T_k^i = \mathbb{1}\{T^i > t_k\} \in \{0, 1\}$ denotes whether T^i exceeds t_k, i.e. whether the patient i survives beyond the time t_k, and $t_1 \prec t_2 \prec \ldots \prec t_{K-1}$. Figure 1 illustrates an example of $K = 6$ time bins and two patient encodings whose events are recorded at t_3 and t_4, respectively.

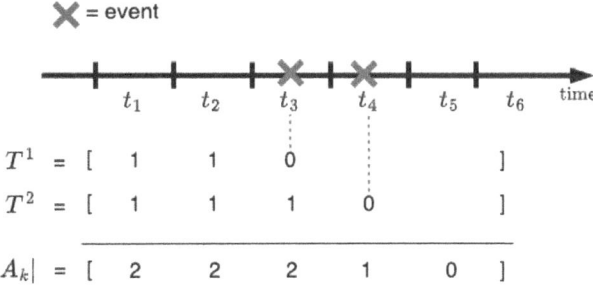

Fig. 1. An illustration of the uncensored encodings of a batch of two patients who experience an event at t_3 and t_4, respectively, from a total of $K = 6$ time bins. Each entry represents $T_k^i = \mathbb{1}\{T^i > t_k\}$. $|A_k|$ is the size of the conditional training subsets for each time bin in the batch.

Censored Encoding ($\delta^i = 0$). For censored patients, only the lower bound of the patients' survival times are known, i.e., that they are alive or did not experience an event at least up until the recorded time t_k. In this case, the actual time-to-events are unknown, and the patients have a chance of experiencing an event at any future time bin. Consider a third patient who is censored at t_3. Assuming independent censoring, where the patient is equally likely to experience an event as others with similar covariates, the patient's encoding can be expanded to account for all possibilities of the event happening from the time of censoring to the final time bin (Fig. 2). In presenting the model with all possible combinations of time-to-events, the model also implicitly learns that a patient's survival probability decreases over time because the ratio of survival to event occurrence also decreases.

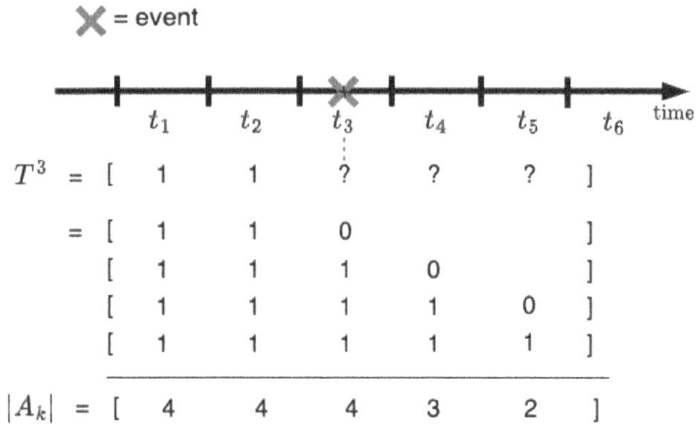

Fig. 2. An illustration of the expanded censored encodings of a batch of one patient whose event information is unknown at t_3 onwards, from a total of $K = 6$ time bins. Each entry represents $T_k^i = \mathbb{1}\{T^i > t_k\}$. $|A_k|$ is the size of the conditional training subsets for each time bin in the batch. Here, the patient's survival probability beyond a time bin t is $\sum A_k / |A_k|$, i.e., 100% (4/4) for t_1 and t_2, indicating certainty in the patient's survival status, followed by 75% (3/4) for t_3, 67% (2/3) for t_4, and 50% (1/2) for t_5.

2.2 Network Output

The output of the network is a sequence of conditional probabilities using conditional training subsets A_k at each time bin where the sigmoid output of each neuron is interpreted as:

$$f_k(X^i) = \sigma(z^i) = P(T^i > t_k | T^i > t_{k-1}, X^i) \tag{1}$$

with nested events $\{T^i > t_k\} \subseteq \{T^i > t_{k-1}\}$, where f_k is the output of the k-th neuron, corresponding to the k-th time bin.

The survival output at each time bin can be computed by applying the chain rule for probabilities to the output neurons:

$$S(T^i > t_k | X^i) = \prod_{j=1}^{k} f_j(X^i) \qquad (2)$$

Since each probability lies between 0 and 1, as the time bin index increases, this cumulative product guarantees a monotonically decreasing function that is desired for survival curves. To obtain the time-bin index of the prediction, we calculate

$$\hat{T}^i = 1 + \sum_{k=1}^{K-1} \mathbb{1}\{f_k(X^i) > 0.5\} \qquad (3)$$

corresponding to the median of the survival curve [20]. To maintain a conditional probability interpretation, we construct the conditional training subsets as proposed by [21] with sizes $|A_k|$ equal to the number of samples in the batch that satisfies $\{T^i > t_k\}$, yielding $|A_1| \geq |A_2| \geq ... \geq |A_{K-1}|$ (see Figs. 1 and 2 for illustration).

2.3 SurvCORN Loss

The SurvCORN loss consists of two components: a log-likelihood and a ranking loss. Following the above, the log-likelihood is

$$loss_{LL} = -\frac{1}{\sum_{k=1}^{K-1}|A_k|} \sum_{k=1}^{K-1} \sum_{i=1}^{|A_k|} [\mathbb{1}\{T^i > t_k\} \cdot log(f_k(X^i)) + \mathbb{1}\{T^i \leq t_k\} \cdot log(1-f_k(X^i))] \qquad (4)$$

The ranking loss directly optimizes the concordant index by penalizing incorrect ordering of pairs and is adapted from [17]:

$$loss_{rank} = \sum_{i,j} \delta^i \mathbb{1}\{T^i < T^j\} \exp\left(\frac{\hat{S}(T^i \mid X^i) - \hat{S}(T^j \mid X^j)}{\alpha}\right) \qquad (5)$$

where α is a hyperparameter set to 0.1 following [17]. The final SurvCORN loss is a summation of the two:

$$\mathcal{L}_{final} = loss_{LL} + loss_{rank} \qquad (6)$$

The log-likelihood drives the model to learn the correct survival times of each patient (i.e., *intra-patient* predictions), while the ranking loss encourages the model to learn the correct ordering between patient survival times (i.e. *inter-patient* predictions).

3 Evaluation Metric

We report the time-dependent concordant index (C-index) of [2] and introduce a new metric called SurvMAE.

The C-index is a typical metric reported in survival analysis [8–10]. It compares the ordering of every pair of patients and quantifies the proportion of pairs in the dataset whose predicted survival times are concordant with the actual survival times. A pair of individuals is considered concordant if the individual with the longer predicted survival time (or conversely, the lower predicted risk) also has the longer observed survival time. As Harrel's C-index [8–10] was originally derived for the proportional hazards framework, Antolini et al. [2] extended it to the non-proportional cases by introducing a time-dependent variation of the C-index which we employ in this paper. The C-index accounts for the *relative* ordering of patients, but does not take into consideration the actual predicted times of a survival model.

To this end, we propose survival mean absolute error (**SurvMAE**), a simple metric built upon MAE that accounts for different censoring mechanisms. Specifically, for censored patients whose actual time-to-events are unknown, the MAE is calculated as the average of the predicted times to the left of (i.e., less than) or equal to the actual times. For uncensored patients, the MAE is calculated normally for predictions to the left or right of the actual times. Mathematically, SurvMAE is represented by the following:

$$SurvMAE = \frac{1}{N_U} \sum_{i=1}^{N_U} \left[\delta^i ||\hat{T}^i - T^i||_1 \right] + \frac{1}{N_C} \sum_{j=1}^{N_C} \left[(1-\delta^i) ||\hat{T}^j - T^j||_1 \cdot \mathbb{1}\{\hat{T}^j \leq T^j\} \right] \tag{7}$$

where N_U is the number of uncensored patients, N_C is the number of censored patients, δ is the event indicator (0 for censored, 1 otherwise), and $\hat{T}$ and T are the predicted and actual survival times, respectively. For continuous-time survival models, SurvMAE gives a directly interpretable prediction error in unit time (i.e., in terms of the number of days, months, etc.). For discrete equidistant time bins, SurvMAE can be translated into an interpretable time-to-event metric through interpolation, where the lower the SurvMAE the better.

Reporting both C-index and SurvMAE is crucial to give a more holistic understanding of the performance of survival models. C-index provides insight regarding the correct ordering of patients, while SurvMAE evaluates the individual predicted time-to-events of the patients.

4 Experimental Setup

4.1 Datasets

We assess the prognostic ability of SurvCORN by comparing it against conventional benchmarks in analyzing two real-world cancer datasets: ChAImeleon [3] and HECKTOR [1,18].

ChAImeleon. [3] is a lung cancer dataset comprised of CT scans and EHR from 320 different patients, with 59% of the data being censored. The average survival time of the patients is 40.5 months with a standard deviation of 20 months. In this work, we employ a segmentation model from [12] to crop the lung areas, allowing the prognostic models to focus on the region of interest.

HECKTOR. [1,18] is a multi-centric, multi-modal head-and-neck cancer dataset consisting of 224 CT and PET scans with corresponding EHRs, with 75% of the data being censored. The average survival time of the patients is 27.8 months with a standard deviation of 24.3 months. The PET and CT scans are registered to a common origin to enable accurate integration of useful information from the two imaging modalities.

We preprocess the images in each dataset to standardize the prognostic model input by performing resampling, cropping, and resizing to a final input size of $112 \times 112 \times 130$. Additionally, we use equidistant binning of the survival time distribution where the number of time bins for each dataset is obtained as the square root of the number of observations.

4.2 Implementation Details

We run all experiments for 30 epochs using a five-fold cross-validation with a DenseNet-121 [13] architecture, a batch size of 16, Adam [15] optimizer with a learning rate of 1×10^{-3} and a weight decay of 1×10^{-5}. We maintain a constant ratio of patients who experienced each event and those who were censored in each fold. All experiments are implemented using PyTorch [19]. Since the survival methods differ primarily in their loss function and treatment of the network output, we keep the network architecture and hyperparameters fixed to ensure a fair and unbiased comparison across datasets and experiments.

5 Results

We compare SurvCORN against three deep baseline survival models (i.e. Nnet-Survival [7], DeepMTLR [6], and DeepHit [17]), all of which treat the survival times as discrete. These models have been shown to achieve competitive overall predictive performance in the survival analysis literature. We report the averages and standard deviations of the C-index and SurvMAE in Tables 1 and 2 for the ChAImeleon Lung Cancer [3] and HECKTOR [1,18] datasets, respectively. Our method achieves competitive performance in C-index while it consistently outperforms other methods in SurvMAE.

Table 1. Average concordance indices and SurvMAE (± standard deviations) on the ChAImeleon Lung Cancer [3] dataset using deep discrete-time survival models. SurvMAE (time) is calculated through a linear interpolation of SurvMAE (bin) where the survival curves hit 0.5. All experiments are run with five-fold cross-validation. The best scores per dataset are bolded.

	C-index ↑	SurvMAE (bin) ↓	SurvMAE (time) ↓
Nnet-Survival [7]	**0.655 ± 0.07**	4.55 ± 0.71	62.7 ± 12.7
DeepHit [17]	0.625 ± 0.06	4.62 ± 1.25	64.0 ± 13.7
SurvCORN (ours)	0.633 ± 0.05	**4.27 ± 0.51**	**52.9 ± 24.9**

Table 2. Average concordance indices and SurvMAE (± standard deviations) on the HECKTOR [1,18] dataset using deep discrete-time survival models. All experiments are run with five-fold cross-validation. SurvMAE (time) is calculated through a linear interpolation of SurvMAE (bin) where the survival curves hit 0.5. The best scores per dataset are bolded.

	C-index ↑	SurvMAE (bin) ↓	SurvMAE (time) ↓
Nnet-Survival [7]	0.683 ± 0.03	6.20 ± 0.50	38.4 ± 10.0
DeepHit [17]	0.674 ± 0.04	6.02 ± 0.25	57.3 ± 58.1
SurvCORN (ours)	**0.705 ± 0.04**	**5.06 ± 0.08**	**32.6 ± 2.5**

6 Discussion and Conclusion

Compared to Nnet-Survival [7], DeepHit [17] performs slightly worse, consistent with the findings in literature (e.g. [16]). For censored data, Nnet-Survival [7] allows the model to operate without penalization beyond the recorded survival time, while SurvCORN implicitly models a decreasing survival probability beyond the recorded survival time. The conditional probability formulation of SurvCORN encourages the model to learn predictions that closely approximate the actual time-to-events, thus optimizing for SurvMAE. Additionally, like DeepHit [17], SurvCORN employs a ranking loss to promote the correct ordering of survival times, thus optimizing for C-index.

Our model, SurvCORN, not only competes effectively with existing baselines in terms of C-index predictions but also demonstrates superior performance in SurvMAE scores. This distinction is particularly noteworthy as the C-index has been deemed unreliable, uninterpretable, and clinically less useful in many works (e.g. [4,11,23]). SurvMAE evaluates the accuracy of the predicted time-to-events of the patients, thus reflecting our model's precision in estimating event occurrences. This holds significance in prognosis outlook prediction as it gives clinicians timely insights to make informed decisions on treatment plans, resource allocation, and patient management strategies.

Limitations. Like other discrete survival methods, SurvCORN does not allow extrapolation or finer-grained ordering of the patients beyond the maximum time-to-event recorded in the train set. Another limitation is that the encoding for censored patients (see *Sect.* 2.1) encourages the model to learn a decreasing survival probability over time beyond the censored time. While this is a natural and reasonable assumption in many cases, it is not always true. It is possible, for example, that a critically ill patient with early censoring may experience an event sooner.

References

1. Andrearczyk, V., et al.: Overview of the HECKTOR challenge at MICCAI 2021: automatic head and neck tumor segmentation and outcome prediction in PET/CT images. In: Andrearczyk, V., Oreiller, V., Hatt, M., Depeursinge, A. (eds.) HECKTOR 2021. LNCS, vol. 13209, pp. 1–37. Springer, Cham (2022). https://doi.org/10.1007/978-3-030-98253-9_1. https://api.semanticscholar.org/CorpusID:245877569
2. Antolini, L., Boracchi, P., Biganzoli, E.: A time-dependent discrimination index for survival data. Stat. Med. **24**(24), 3927–3944 (2005)
3. Bonmatí, L.M., et al.: CHAIMELEON project: creation of a Pan-European repository of health imaging data for the development of AI-powered cancer management tools. Front. Oncol. **12** (2022)
4. Cook, N.R.: Use and misuse of the receiver operating characteristic curve in risk prediction. Circulation **115**, 928–935 (2007). https://api.semanticscholar.org/CorpusID:14594808
5. Cox, D.R.: Regression models and life-tables. J. Roy. Stat. Soc. Ser. B (Methodol.) **34** (1972). https://doi.org/10.1111/j.2517-6161.1972.tb00899.x
6. Fotso, S.: Deep neural networks for survival analysis based on a multi-task framework. ArXiv abs/1801.05512 (2018). https://api.semanticscholar.org/CorpusID:13482950
7. Gensheimer, M.F., Narasimhan, B.: A scalable discrete-time survival model for neural networks. PeerJ **7**, e6257 (2019)
8. Harrell, F.E., Califf, R.M., Pryor, D.B., Lee, K.L., Rosati, R.A.: Evaluating the yield of medical tests. JAMA J. Am. Med. Assoc. **247** (1982). https://doi.org/10.1001/jama.1982.03320430047030
9. Harrell, F.E., Lee, K.L., Califf, R.M., Pryor, D.B., Rosati, R.A.: Regression modelling strategies for improved prognostic prediction. Stat. Med. **3** (1984). https://doi.org/10.1002/sim.4780030207
10. Harrell, F.E., Lee, K.L., Mark, D.B.: Multivariable prognostic models: issues in developing models, evaluating assumptions and adequacy, and measuring and reducing errors. Stat. Med. **15** (1996). https://doi.org/10.1002/(SICI)1097-0258(19960229)15:4⟨361::AID-SIM168⟩3.0.CO;2-4
11. Hartman, N., Kim, S., He, K., Kalbfleisch, J.D.: Pitfalls of the concordance index for survival outcomes. Stat. Med. **42** (2023). https://doi.org/10.1002/sim.9717
12. Hofmanninger, J., Prayer, F., Pan, J., Röhrich, S., Prosch, H., Langs, G.: Automatic lung segmentation in routine imaging is primarily a data diversity problem, not a methodology problem. Eur. Radiol. Exp. **4**(1), 50 (2020)
13. Huang, G., Liu, Z., Van Der Maaten, L., Weinberger, K.Q.: Densely connected convolutional networks. In: 2017 IEEE Conference on Computer Vision and Pattern Recognition (CVPR), pp. 2261–2269 (2017). https://doi.org/10.1109/CVPR.2017.243

14. Katzman, J.L., Shaham, U., Cloninger, A., Bates, J., Jiang, T., Kluger, Y.: DeepSurv: personalized treatment recommender system using a cox proportional hazards deep neural network. BMC Med. Res. Methodol. **18** (2018). https://doi.org/10.1186/s12874-018-0482-1
15. Kingma, D., Ba, J.: Adam: a method for stochastic optimization. In: International Conference on Learning Representations (ICLR), San Diega, CA, USA (2015)
16. Kvamme, H., Borgan, Ø.: Continuous and discrete-time survival prediction with neural networks. Lifetime Data Anal. **27**(4), 710–736 (2021). https://doi.org/10.1007/s10985-021-09532-6
17. Lee, C., Zame, W., Yoon, J., Van der Schaar, M.: DeepHit: a deep learning approach to survival analysis with competing risks. Proc. Conf. AAAI Artif. Intell. **32**(1) (2018)
18. Oreiller, V., et al.: Head and neck tumor segmentation in PET/CT: the HECKTOR challenge. Med. Image Anal. **77**, 102336 (2022). https://doi.org/10.1016/j.media.2021.102336
19. Paszke, A., et al.: PyTorch: an imperative style, high-performance deep learning library. In: Advances in Neural Information Processing Systems, vol. 32 (2019)
20. Reid, N.: Estimating the median survival time. Biometrika **68** (1981). https://doi.org/10.1093/biomet/68.3.601
21. Shi, X., Cao, W., Raschka, S.: Deep neural networks for rank-consistent ordinal regression based on conditional probabilities. Pattern Anal. Appl. **26** (2023). https://doi.org/10.1007/s10044-023-01181-9
22. Sparr, L.F., Moffitt, M.C., Ward, M.F.: Who returns and who stays away. Am. J. Psychiatry **150**, 801–805 (1993)
23. Vickers, A.J., Cronin, A.M.: Traditional statistical methods for evaluating prediction models are uninformative as to clinical value: towards a decision analytic framework. Semin. Oncol. **37** (2010). https://doi.org/10.1053/j.seminoncol.2009.12.004

Author Index

A
A.E. Hellström, Terese 3
Aaron Gulliver, T. 59
Abdi, Abas 124
Abouassaly, Robert 167
Acosta, Oscar 211
Adams-Tew, Samuel 190
Ahmadvand, Noushin 124
Ali, Mansoor 59
Amer, Alyaa 124
Andratschke, Nicolaus 221
Avendano Avalos, Daly Betzabeth 59
Azam, Muhammad Adeel 133
Azarmehr, Neda 124

B
Bagci, Ulas 26
Baldini, Chiara 133
Barros, Vesna 167
Bartholomew, Angelica 167
Bergman, Jacques J. 83
Bert, Julien 114
Bhusal, Bhumi 26
Bodenmann, Tobias R. 37
Boers, Tim G. W. 83
Bône, Alexandre 104
Bosques Palomo, Beatriz Alejandra 59
Brahmavar, Shreyas Bhat 37
Bridge, Christopher P. 37
Briens, Aurélien 211
Bruining, Annemarie 3

C
Campbell, Rebecca 167
Campbell, Steven C. 167
Cardona-Huerta, Servando 59
Chourak, Hilda 211
Cleveland, Mason C. 37
Colantonio, Sara 48

D
Dang, Hong-Phuong 114
Day, Lorena 190
de Crevoisier, Renaud 211
de Groof, Albert J. 83
De With, Peter H. N. 83
Decaens, Thomas 104
Diba, Ali 154
Dillard, Laurent 154
Dorfner, Felix J. 37

E
Eftestøl, Trygve 73
Eklund, Martin 73
Elliott, Jessie A. 176
Ermis, Ekin 221

F
Fernandez-Quilez, Alvaro 73

G
Gabrani, Maria 167
Garza Abdala, Jorge Alberto 59
Gerstner, Elizabeth 37
Ghatwary, Noha 124
Gil, Nelson 37
Glaunes, Joan Alexis 104
Goebel, Clara 167
Guelen, Melisa S. 37
Gutiérrez-Becker, Benjamín 93

H
H.B. Claessens, Cris 3
H.N. De With, Peter 3
Heller, Nicholas 167
Hong, Chenyi 14
Hussain, Sadam 59
Hussein, Alaa 124

I
Ioppi, Alessandro 133

J
Jäderling, Fredrik 73
Jaspers, Tim J. M. 83
Jha, Debesh 26
Jia, Dengqiang 200
Johnson, Audrey 190
Johnson, Sara 190
Jong, Martijn R. 83
Joshi, Sarang 190

K
Kalpathy-Cramer, Jayashree 37
Kamath, Amith 221
Khan, Ufaq 144
Kim, Albert E. 37
Kim, Tae Soo 154
Koch, Anna 3
Kooi, Thijs 154
Kotowski, Krzysztof 93
Krason, Agata 93
Kucharski, Damian 93
Kusters, Carolus H. J. 83

L
Lee, Hyeonsoo 154
Lee, Weonsuk 154
Liu, Zonglin 200
Liu, Zuozhu 14
Lv, Junhui 14

M
M.J. Piek, Jurgen 3
Ma, Qiong 200
Maani, Fadillah Adamsyah 231
Machura, Bartosz 93
Magdy, Sahar 124
Malavaud, Bernard 114
Markar, Sheraz R. 176
Mattos, Leonardo S. 133
Meng, Runqi 200
Mercado, Zahira 221
Miranda, Gustavo Andrade 114
Monsivais Molina, Mario Alexis 59
Morato, Beatriz López 167
Mose, Lucas 221

N
Nalepa, Jakub 93
Nandakumar, Karthik 231
Naseem, Usman 59
Nawaz, Umair 144
Nederend, Joost 3
Niers-Stobbe, Ilse 3
Nies, Ingrid 3
Nordström, Tobias 73

O
Odeen, Henrik 190
Ozery-Flato, Michal 167

P
Pachetti, Eva 48
Papanikolopoulos, Nikolaos 167
Papież, Bartłomiej W. 176
Patel, Jay B. 37
Payne, Allison 190
Peretti, Giorgio 133
Pessin, Alissa 190
Poel, Robert 221
Pulido-Arias, Dagoberto 37

R
Ramesh, Vidhyalakshmi 167
Regmi, Nisha 26
Regmi, Smriti 26
Regmi, Subodh 167
Remer, Erick 167
Reyes, Mauricio 221
Reynolds, John V. 176
Ridzuan, Muhammad 231
Riggenbach, Elena 221
Rosen, Bruce R. 37
Rosen-Zvi, Michal 167

S
Saddik, Abdulmotaleb E. 144
Saeed, Numan 231
Sampieri, Claudio 133
Scovell, Jason 167
Septiers, Valentin 211
Serej, Nasim Dadashi 124
Seshadri, Rikhil 167

Author Index

Shea, Jill 190
Shen, Dinggang 200
Singh, Amanpreet 190
Siva, Jayant 167
Sortland Rolfsnes, Erlend 73
Sosa-Marrero, Carlos 211
study group, ENSURE 176
Subedi, Aliza 26
Sun, Kaicong 200
Sun, Yiqun 200

T

Taguelmimt, Kamilia 114
Tamez Pena, Jose Gerardo 59
Tejpaul, Resha 167
Teng, Lin 200
Tessier, Jean 93
Thangngat, Philip 73
Thiran, Jean-Philippe 37
Thorniley, Madelaine 133
Tong, Tong 200

V

van der Sommen, Fons 3, 83
van Eijck van Heslinga, Rixta A. H. 83
Visvikis, Dimitris 114

W

W.R. Schultz, Eloy 3
Wallerstein-King, Gabriel 167
Wang, Hualiang 14
Ward, Ryan 167
Weight, Christopher 167
Willmann, Jonas 221
Wood, Andrew 167
Wu, Zhuoxuan 14

Y

Yang, Sisi 104
Yaqub, Mohammad 231

Z

Zheng, Yuhan 176
Zuluaga, Maria A. 211

SPRINGER NATURE

GPSR Compliance

The European Union's (EU) General Product Safety Regulation (GPSR) is a set of rules that requires consumer products to be safe and our obligations to ensure this.

If you have any concerns about our products, you can contact us on ProductSafety@springernature.com

In case Publisher is established outside the EU, the EU authorized representative is:

Springer Nature Customer Service Center GmbH
Europaplatz 3
69115 Heidelberg, Germany

The manufacturer's authorised representative in the EU is Springer Nature Customer Service Centre GmbH, Europaplatz 3, 69115 Heidelberg, Germany. If you have any concerns regarding our products, please contact ProductSafety@springernature.com

Printed and bound by CPI Group (UK) Ltd, Croydon, CR0 4YY

26/03/2026

02078933-0013